D0951865

GALILEO'S MIDDLE FINGER

GALILEO'S MIDDLE FINGER

HERETICS, ACTIVISTS, AND THE SEARCH FOR JUSTICE IN SCIENCE

ALICE DREGER

PENGUIN PRESS
New York
2015

PENGUIN PRESS
Published by the Penguin Group
Penguin Group (USA) LLC
375 Hudson Street
New York, New York 10014

USA • Canada • UK • Ireland • Australia •
New Zealand • India • South Africa • China

penguin.com
A Penguin Random House Company

First published by Penguin Press,
a member of Penguin Group (USA) LLC, 2015

LIBRARY OF CONGRESS CATALOGING-IN-PUBLICATION DATA
Dreger, Alice Domurat, author.
Galileo's middle finger: heretics, activists,
and the search for justice in science / Alice Dreger.
pages cm
Includes bibliographical references and index.
ISBN 978-1-59420-608-5
1. Science—Moral and ethical aspects. 2. Science—Political aspects. 3. Scientists—Professional ethics.
4. Heresy in science. I. Title.
Q175.35.D74 2015
174.2'8—dc23
2014036659

Printed in the United States of America
1 2 3 4 5 6 7 8 9 10

Designed by Marysarah Quinn

FOR KEPLER, who saved his mother.

CONTENTS

GALILEO'S

MIDDLE

FINGER

INTRODUCTION

THE TALISMAN

SOON ENOUGH, I will get to the death threats, the sex charges, the alleged genocides, the epidemics, the alien abductees, the antilesbian drug, the unethical ethicists, the fight with Martina Navratilova, and of course, Galileo's middle finger. But first I have to tell you a little bit about how I got into this mess. And explain why I think we now have a very dangerous situation on our hands.

As an academic historian who typically hangs out with her own political kind, I'm aware of the stereotype many liberals have about conservative Catholics. The former believe the latter don't think—that conservative religious people don't care about facts and rigorous inquiry. But my conservative Catholic parents *were* thinkers. Twice as often as my parents told their four children to go wash, they told us to go look something up. At our suburban tract house on Long Island in the 1970s, our parents shelved the *Encyclopædia Britannica* right next to the dinner table so we could easily reach for a volume to settle the frequent debates. The rotating stack of periodicals in our kitchen included not only religiously oriented newsletters, but also the *New York Times* and *National Geographic*. Our parents took us to science museums, woke us up for lunar eclipses, and pushed us to question our textbooks and even our teachers when they sounded wrong. Although our mother never mentioned that she had earned a degree in philosophy from Hunter College, she read to us aloud from Plato and

Shakespeare, analyzing the texts as she read. Meanwhile, our father, a draftsman for one of the big Long Island defense contractors, loved learning in spite of having had only a high school education. We joked that he would someday be crushed under his books, most of them military histories of Poland, the homeland of both sides of our family. He got us microscopes and telescopes and talked seriously about the potential for alien life-forms. I vividly recall that, when one day we summoned him urgently to come see a giant UFO that had appeared in the sky, he was genuinely disappointed to discover he had bothered to grab his camera for the Goodyear blimp.

But besides being intellectuals and knowledge seekers, my parents were also industrial-strength Roman Catholics. They sought out Latin masses and avoided meat on Fridays long after Vatican II declared all that fuss unnecessary. They sent us to public school not only because the local public schools offered the best education around, but also because the local Catholic school struck them as dangerously liberal in its religious orientation. (Better to be among Protestants and Jews than roomfuls of squishy Catholics.) Their religious devotion manifested itself largely in pro-life activism. Even while their own children were still young and underfoot, my parents collected baby things to give to poor mothers, took in a young pregnant woman who had been thrown out by her parents, and became foster parents to a mixed-race baby of a single mother, ultimately adopting that child. As we were growing up, the basement of our house slowly filled with homemade placards we would carry when marching outside abortion clinics.

Although they were highly obedient to authority in their religious lives, in their political lives, my parents were rabble-rousers. My father ran for Congress on the Right-to-Life Party line, while my mother helped lead the local chapter of Feminists for Life. (In the 1970s, bra-burning pro-lifers were a real thing.) My mother especially embraced her American rights to speak, to assemble, to vote, and to protest, because she knew her life might well have turned out differently. Born in 1935 in Poland, she had somehow survived the Second World War with her extended family in their tiny farming village in an area subjected to repeated aerial bombings and ground-war skirmishes. Not long after the war ended, at the age of eleven, she had been suddenly transported with her brother and mother to America, where the three of them were reunited with her father. (Her father had had dual citizenship and had fought with the Americans.) On these shores, she found a land where you could, *without fear*, say and think what you wanted, worship and vote as you wanted, and openly object to what you found stupid or offensive. She let us know, as we were growing up, that she considered American democracy a true wonder, a tool to be used at every chance. The Bill of Rights seemed to her almost as sacred as the Bible. This view was implicitly and explicitly reinforced by the rare relatives who made it out of Soviet-controlled Poland and came to lodge with us.

My parents never seemed to feel a tension between these heavy strands that comprised their lives—the Old World and the New, the religious and the intellectual, the obedient and the activist. I suppose that to them it all seemed obviously interrelated. They had no trouble sending me to confession one day and renewing my subscription to *Natural History* magazine the next. But as I grew up, I felt the tension one surely *must* feel when being simultaneously taught the importance of a specific dogma and the importance of freedom *from* dogma.

I knew that some people abandoned their parents' religion as a way of asserting their independence. But for me, losing my religion wasn't about rebellion against my parents; indeed, I felt quite forlorn at the idea of disappointing my family by admitting my atheism. Still, my parents' religious faith seemed to me incommensurate with our deeply felt faith in America—a faith in freedom of inquiry, in freedom of thought, in the will and right of the people to collectively discover truth and to make their own rules accordingly. And I loved America much more than I loved the Vatican, that place where celibate old men had the right to tell intelligent women what we should think and do. By the time I was in my late teens, while my sister was on her way to becoming a nun, I couldn't help but notice that the place I felt the hope of salvation wasn't church. It was the American Museum of Natural History, that great cathedral of evolution. As often as I could, I would take the train into New York City and lie under the giant blue whale in the great darkened hall of ocean life. Every time I lay there—waiting for the delicious moment when the whale started to move, from optical illusion—science struck me as the obvious and perhaps only way to remain perpetually free from blinding, oppressive dogma.

I guess, then, it is not too surprising that I ultimately decided to pursue a PhD in the history and philosophy of science, at Indiana University. Exploring the very life and guts of science by studying the history and the philosophy of it—*this* seemed to me the way to make sure that the most antidogmatic way of life we had available to us, the scientific way of life, would remain healthy and vigorous. But by the time I moved to Bloomington for graduate school, in 1990, not everyone in the academic fields of science studies (the history, philosophy, and sociology of science) felt the same devotion. At that point, Marxist and feminist science-studies scholars had for almost two decades been producing a large body of work deeply critical of various scientific claims and practices. They had shown how various scientists had, in word and deed, oppressed women, people of color, and poor folks, typically by making problematic "scientific" claims about them. Harvard biologist Ruth Hubbard, for example, had taken apart pseudoscientific claims that biology made women "naturally" less capable of doing science than men. Historians like Londa Schiebinger and Cynthia Eagle Russett had documented how, over many centuries, patriarchies had deployed the rhetoric of science to represent women as inherently inferior to men. Meanwhile, Hubbard's Harvard colleague Stephen Jay Gould had scrutinized "scientific" studies purporting to show important racial differences in skull size and IQ and had shown them to be hopelessly riddled with racist bias.

Make no mistake: As a liberal feminist, I *was* extremely sympathetic to feminist and Marxist science studies. Indeed, the work of scholars like Gould—whose columns in *Natural History* I had devoured as a teenager—struck me as constituting perhaps the most important work of social justice of our time, because it challenged racist and sexist claims about human nature. These leftist criticisms were part of what drove me to graduate school. But to me at least, the finding by Gould and others that scientists often suffered from bias didn't mean science *itself* was rotten. The very fact that scholars could *see and show* problems of racist and sexist bias in science stood to me as proof that, together, evidence-driven scholars could advance knowledge and ultimately get past the individual human mind's tendency to follow familiar scripts. If some of the products of science disappointed me, the process most assuredly did not. Indeed, in graduate school, I gravitated toward historical work specifically because I loved the relatively scientific process in history of seeking, organizing, and analyzing evidence—of letting the data guide you toward new and unexpected learning, as much as humanly possible.

IN GRADUATE SCHOOL, I ended up cutting my scholarly teeth on the history of the biomedical treatment of people born with sex anomalies—the people who used to be called hermaphrodites. For many years, people would assume I had a personal stake in this identity issue—that I or someone I loved had been born hermaphroditic—but in fact this topic was simply suggested to me by my dissertation director, who saw it as a great way to examine "scientific" conceptions of gender, something that fascinated me as a feminist. To be honest, in looking into the history of hermaphroditism, I decided to focus on the late nineteenth and early twentieth centuries because I figured I'd find easy pickings there. I already knew that most doctors of that time were politically conservative men, inclined to believe that the unequal social treatment of women arose from—nay, was *required* by—the allegedly natural two-sex divide. I knew there would have been a lot at stake for one of these sexist doctors when a patient appeared on inspection to be a hermaphrodite. Some of these patients had immediately apparent mixes of male and female traits—a notable phallus and a vaginal opening or feminine breasts along with a full beard. Others appeared to have one sex externally but the opposite internally. All unwittingly challenged the idea that there were only two real sexes—that there was a clear, natural divide between men and women.

Just as I was finishing my PhD, in 1995, I published my first scholarly paper, in the journal *Victorian Studies*. This article mapped out a hitherto uncharted history: what Victorian British doctors had done when faced with living proof that humans don't come in only two sexes. Though my report contained some grainy 1890s photographs of ambiguous genitalia, it was still pretty academic, showing no real hint of the odd path the paper's publication would lead me down. My finding was simply that Victorian doctors, befuddled by cases of "doubtful sex," had deployed pragmatic combinations of clever rhetorical strategies, new scientific tools like microscopes, and the occasional surgical scalpel to try to make "true hermaphroditism" virtually disappear, all to protect long-standing social distinctions between men and women. But dry as that article may have been, it ended up pushing me into two unfamiliar and intense worlds: contemporary sex politics and contemporary medical activism. That's because, thanks to the Internet, by the time I came to this topic, in the mid-1990s, something was going on that the Victorian doctors would never have imagined: People who had been born with various sex anomalies had started to find each other, and they had started to organize as an identity rights movement.

Labeling themselves *intersex*, many gathered under the leadership of Bo Laurent, the founder of the Intersex Society of North America, and after reading my *Victorian Studies* article, some of these intersex activists, including Bo, contacted me. A couple wrote me simply to complain that they found some of my language offensive, apparently not realizing I was relaying Victorian rhetoric in my article. By contrast, Bo got my work. And she asked for my help in changing the way children born intersex were treated in modern medicine.

Now, as a straight, sex-typical female earning degrees in history and philosophy, I had started working in this field not only rather uneducated about human sex anatomy, but also rather uneducated about the politics of contemporary medicine. Still, it didn't take long for me to see the ways that our present-day medical system was indeed as broken as Bo and her compatriots were describing. Indeed, the system being employed at the children's hospital down the street from my grad-school apartment made the Victorian approach look relatively benign. The modern system featured not only highly aggressive cosmetic genital surgeries in infancy for children born with "socially inappropriate" genital variations like big clitorises, but also the withholding of diagnoses from patients and parents out of fear that they couldn't handle the truth. It treated boys born with small penises as hopeless cases who "had" to be castrated and sex-changed into girls, and it assumed that the ultimate ability of girls to reproduce as mothers should take precedence over all else, including the ability to someday experience orgasm.

I hastened to tell Bo, "I'm a historian; I study *dead* people." However, once I understood what was really going on at pediatric hospitals all over the nation—once I understood that Bo's clitoris had been amputated in the name of sex "normalcy" and that this practice was still going on—I felt I had to assist in her efforts. I had been raised to be an activist and to be someone who helps people in desperate circumstances, and I was stunned and outraged by what was going on. I threw myself into the struggle and spent the decade after grad school living two lives—as a professor researching and writing academic histories of the medical establishment's treatment of intersex and also as a patient advocate and a leading activist for the rights of sexual minorities. By day, I was your typical history professor—researching, teaching, and dealing with committee assignments. By night, I was campaigning to stop unnecessary and harmful genital surgeries, ill-advised sex changes on babies, and the well-meaning lies told to affected families. I held fund-raisers, I drafted press releases, I developed policies, I wrote and ghost-wrote propaganda, and I stuffed a lot of envelopes. I also testified to governmental committees, met with groups of activists and doctors, got media training, and appeared as a talking head on one news program after another.

I found the advocacy work so meaningful and so exhausting that when it was time for me to go up for promotion to full professorship, I quit my day job instead. About ten years into my life as a PhD, I gave up tenure and the ability to grow my retirement account in part so that I would have more time and energy for activism. I also did it because by then I'd had a kid and couldn't continue to devote myself to two jobs; until I turned in my resignation letter, on top of my job as a professor, I was also managing our staff of five at the Intersex Society. At that point, I did let an old academic friend talk me into picking up a part-time, untenured professorship at Northwestern University's medical school in Chicago. The job there was small enough to leave me free to do whatever I felt needed my attention but big enough in name to open some doors.

IT WAS SHORTLY AFTER this time that I took on a new scholarly project, one that without much warning forced me to question my politics and my political loyalties, if not also my decision to give up tenure. This was a project that suddenly changed me from an activist going after establishment scientists into an aide-de-camp to scientists who found themselves the target of activists like me. Indeed, this project soon put me in a position I would never have imagined for myself: vilified by gender activists at the National Women's Studies Association meeting and then celebrated at the Human Behavior and Evolution Society by the enemies of my childhood hero, Stephen Jay Gould.

The scholarly project, which I took on early in 2006, involved investigating the history of one particular controversy over transgender. Just to be clear, although both transgender and intersex people are historically oppressed sexual minorities, transgender is different from intersex. Whereas *intersex* refers to the condition of being born with a mix of female and male anatomical features, being *transgender* means feeling that the gender label assigned to you at birth was the wrong one. Think Christine Jorgensen or Chaz Bono, people who were born clearly one sex but who find they need to change it. To oversimplify it a bit, we could say that intersex is primarily about how you are born in terms of your sex organs, and transgender is primarily about how you feel in terms of your gender identity.

In 2003, three years before I came to the story, a group of transgender activists had kicked up a storm over a book by a Northwestern sex researcher, J. Michael Bailey, because in that book, Bailey had pushed a theory these activists didn't like: Bailey had suggested that, in cases of men who become women, transgender isn't just about gender identity, but also about sexual orientation—about eroticism. This, I already knew, was a no-no among certain groups of transgender activists who insisted that virtually all transgender people are born with the brain of one sex and the body of the other—that transgender identity is just about core inborn gender, not about erotic feelings. To opine about sexual orientation in conjunction with transgender the way Bailey did was to skip into a minefield created by four decades of intense social and medical battles over the nature of transgender identity.

Still, I thought I knew from my background in science studies and a decade of intersex work how to navigate an identity politics minefield, so I wasn't that worried when in 2006 I set out to investigate the history of what had really happened with Bailey and his critics. My investigation ballooned into a year of intensive research and a fifty-thousand-word peer-reviewed scholarly account of the controversy. And the results shocked me. Letting the data lead me, I uncovered a story that upended the simple narrative of power and oppression to which we leftist science studies scholars had become accustomed.

I found that, in the Bailey case, a small group had tried to bury a politically challenging scientific theory by killing the messenger. In the process of doing so, these critics, rather than restrict themselves to the argument over the ideas, had charged Bailey with a whole host of serious crimes, including abusing the rights of subjects, having sex with a transsexual research subject, and making up data. The individuals making these charges—a trio of powerful transgender women, two of them situated in the safe house of liberal academia—had nearly ruined Bailey's reputation and his life. To do so, they had used some of the tactics we had used in the intersex rights movement: blanketing the Web to make sure they set the terms of debate, reaching out to politically sympathetic reporters to get the story into the press, doling out fresh information and new characters at a steady pace to keep the story in the media and to keep the pressure on, and rhetorically tapping into parallel left-leaning stories to make casual bystanders "get it" and care. Tracking their chosen techniques was occasionally like reading a how-to activist manual that I could have written, but there was one crucial difference: What they claimed about Bailey simply wasn't true.

You can probably guess what happens when you expose the unseemly deeds of people who fight dirty, particularly when you publish a meticulously documented journal article detailing exactly what they did, and especially when the *New York Times* covers what you found. Certainly I should have known what was coming—after all, I had literally written what amounted to a book on what this small group of activists had done to Bailey. But it was still pretty uncomfortable when I became the new target of their precise and unrelenting attacks. The online story soon morphed into "Alice Dreger versus the rights of sexual minorities," and no matter how hard I tried to point people back to documentation of the truth, facts just didn't seem to matter.

Troubled and confused by this ordeal, in 2008 I purposefully set out on a journey—or rather a series of journeys—that ended up lasting six years. During this time, I moved back and forth between camps of activists and camps of scientists, to try to understand what happens—and to figure out what *should* happen—when activists and scholars find themselves in conflict over critical matters of human identity. This book is the result.

I understand that some people on an exploration like this might have tried to just clinically observe it all and to write an "objective" third-person account of scientific controversies over human identity in the Internet age. But already by the time I set out, I knew way too much about individual human bias to kid myself into thinking I could work simply as a stateless reporter above all the frays. I also felt too strongly the need to honor both good science and good activism to remain uninvolved when I saw crazy stuff happening on one side or the other. I believed—and still believe—too much in the importance of facts to sit idly by when I saw someone, be it a scientist or an activist, actively misrepresenting what is really known. As a consequence, as I traveled through scientist-activist wars over human identity—first in psychology, then anthropology, then prenatal pharmacology—rather than being merely embedded, I kept getting uncomfortably embroiled.

In spite of how difficult some of it has been, this journey of discovery proves something really important: Science and social justice require each other to be healthy, and both are critically important to human freedom. Without a just system, you cannot be free to do science, including science designed to better understand human identity; without science, and especially scientific understandings of human behaviors, you cannot know how to create a sustainably just system. As a consequence of this trip, I have come to understand that the pursuit of evidence is probably the most pressing moral imperative of our time. All of our work as scholars, activists, and citizens of democracy depends on it. Yet it seems that, especially where questions of human identity are concerned, we've built up a system in which scientists and social justice advocates are fighting in ways that poison the soil on which both depend. It's high time we think about this mess we've created, about what we're doing to each other and to democracy itself.

VERY OFTEN DURING this long, strange trip, while stuck in some airport on a layover, I found myself meditating on the image of Galileo's middle finger. I accidentally came upon that mummified digit two decades ago on a trip in graduate school, just at the start of my scholarly work on the history of hermaphrodites. In May 1993, I had gone to Italy to accompany my mother, at her request, on a tour of Roman Catholic religious sites. As we had planned, when the tour ended, my mother flew back to America while I set off to continue around Europe by train to supplement my studies. For my first stop, I took the train from Rome to Florence to visit the history of science museum attached to the Uffizi art galleries. I had planned this short stop in Florence because of the opportunity to see the museum's collection of eighteenth-century wax obstetrical models, life-size teaching instruments I had already read much about. But I was also very excited at the prospect of seeing a set of artifacts that are to a historian of science what Jesus's cross would be to a Christian: Galileo's telescopes.

When Galileo Galilei was born, in 1564, the world had just started changing in the direction that would ultimately lead to modern science, modern technology, and democracy. The old way—accepting authorities without much question—had just started to develop serious cracks. Not long before Galileo's birth, European anatomists like Andreas Vesalius had begun to dissect human bodies and to show that the innards didn't always match what the ancient authorities like Galen described. A Polish scholar named Nicolaus Copernicus had crunched the astronomical numbers and in 1543 suggested a model contrary to the ancient astronomer Ptolemy's, a new model wherein the Sun, not the Earth, formed the center of our world.

But Galileo went much further than these men before him. Philosophically paving the way for the world as we now know it, Galileo actively argued for a bold new way of knowing, openly insisting that what mattered was not what the authorities—ancient or otherwise—*said* was true but what anyone with the right tools could *show* was true. As no one before him had, he made the case for modern science—for finding truth together through the quest for facts.

Galileo's radical new way of thinking (along with his sense of humor) finds perfect display in one particular argument he had with a colleague over a vital and timeless question from the physical sciences: whether you could cook a bird's egg by whirling it around your head in a sling. This hypothetical problem represented a larger physics question about whether flying objects heat up or cool down, but Galileo turned it into the even bigger question: How do we know if something is true?

Galileo's contemporary debate partner on this topic was a Jesuit scientist named Orazio Grassi. Like most people of his time, Grassi usually accepted the word of the ancient authorities, and because ancient authorities reported that the Babylonians had managed to cook eggs by twirling them about in a sling, Grassi figured it must be true. But Galileo mocked this silly claim and in so doing explained how one could personally *test* ideas about cause and effect by controlling for variables, a brilliant and remarkably modern idea. Weighing in on the problem, Galileo wrote:

> If we do not achieve an effect [like cooking an egg by whirling it] which others formerly achieved, it must be that in our operations we lack something which was the cause of this effect succeeding [for our predecessors], and if we lack but one single thing, then this alone can be the cause. Now we do not lack eggs, or slings, or sturdy fellows to whirl them; and still they do not cook, but rather they cool down faster if hot. And since nothing is lacking to us except being Babylonians, then being Babylonians [must be] the cause of the eggs hardening.

Of course what Galileo really meant was not that Babylonians had magical culinary skills, but this: Stop thinking that the authorities know what they're talking about when they're talking about natural causes and effects. Focus your mind on discoverable evidence.

Treating discernable facts as the ultimate authority, Galileo took to doing real experiments, dropping heavy balls down inclined planes to study relative rates of fall, using careful quantification to find predictable, natural patterns in the world. When learned people around Galileo doubted Copernicus's idea that the earth is spinning and racing about the sun—because surely, if we were on a moving, turning planet, everything not tied down would be flying about—Galileo encouraged them to think harder. What happens, he asked them, when you drop a solid object while you are on a moving ship? The object falls straight *down* relative to you and the ship. He encouraged people to see this as real-life analogical *evidence* that could explain why a table not tied down moves *with* the earth's movement and does not fly off. He encouraged them to think beyond the taught or the "obvious," to *see for themselves* what was true.

In the spring of 1609, while living the life of a frustrated, underpaid university professor, Galileo heard about a brand-new optical device, the telescope. Ever the self-starter, he soon constructed one—and then a better one, and a better one. Others saw in this device military and commercial uses. (Ascertaining which trading ships were arriving when could provide advance knowledge of the markets.) But Galileo, engaging his radical epistemology of nature, turned his telescopes to the sky. And what did he see? Not at all the perfect geocentric heavens as they were described by the ancients and taught at the universities. No, indeed. The earth's moon had mountains. (A sign of imperfection in the heavens.) Jupiter had its own moons. (A sign that not everything orbited around the earth.) Venus had phases. (A sign of heliocentrism.) Throughout the sky, Galileo saw evidence of Copernicus's radical new astronomical model.

Unafraid of these new facts and ever confident in his own genius, Galileo didn't even try to reconcile his findings with what the ancients had said. Instead he boldly reported his discoveries in a book he called *The Starry Messenger.* In it, he made a point of including careful drawings to show what the reader could verify with his own eyes if he could get his hands on a decent telescope.

Tempting as it is to see Galileo as supernatural, his surviving writings and the writings of those who knew him personally confirm his humanity for us; they paint him as alternately politically savvy and politically foolish, rash, self-destructive, funny, determined, and devoted to those he loved. And he was deeply human in one other important way: His science was almost certainly motivated, at least in part, by his personal beliefs. The mythical tales of Galileo told by artists like Bertolt Brecht hold him up as a scientific saint, someone who could see completely beyond himself. But as the biographer David Wootton has argued convincingly, Galileo was driven to defend Copernicanism in part because it satisfied his personal psychology: "If Galileo stuck with Copernicanism as the key topic he wanted to write about, it was because he was attracted by the idea of making human beings seem insignificant."

In the hands of Galileo, the telescope became a tool to investigate not only the stars, but also the human condition. He described a messy universe in which we humans are on just another whizzing planet—not a special, still place made for us by an attentive biblical God—and thus strayed dangerously close to the sorts of heretical ideas that had gotten his contemporary Giordano Bruno convicted of heresy. Bruno ended up burned at the stake for putting forth bold new visions of the universe. But Galileo—in spite of repeatedly attracting the attentions of the Inquisition, in spite of being legitimately scared of being subjected to imprisonment and torture and more—could not seem to stop himself from pursuing Copernicanism, from pursuing what he saw must be true about our vast universe, and especially about the rather negligible place of us humans in it. Moreover, he couldn't stop himself from promoting scientific truth in risky ways, even by making the pope look foolish.

This period is often considered the beginning of the Scientific Revolution, but you can see why that term doesn't really capture what Vesalius, Copernicus, Bruno, and especially Galileo were doing. What they were doing was much more radical: This was a revolution in *human identity*. This was not only a shift in ideas about what we can know about the universe, but fundamentally a shift in what we can know *about ourselves.* This was a journey toward what finally became the Enlightenment. When Galileo rejected the Vatican's astronomical dogma, he wasn't rejecting only their "facts" about our planet and our sun; he was also rejecting the Church's right to tell us who we are. There's no doubt: The inquisitors were spot-on to see Galileo as extremely dangerous.

Nevertheless, although the Inquisition could arrest Galileo, it could not arrest human progress. The Scientific Revolution that swept through Europe was soon followed by democratic revolution. And all of these massive changes in science and in politics depended on a single central idea, one that Galileo held dear, the central idea of the Enlightenment: that we get to know for ourselves who we are, by seeking evidence, using reason, and coming to thoughtful consensus on truth. Science and democracy grew up together in Europe and North America, as twins; it is no coincidence that so many of America's Founding Fathers were science geeks. The "American" freedoms to think, to know, to learn, to speak—these were the freedoms that the radical Galileo had seized, long before they were finally written into our laws. As much as Thomas Jefferson and John Adams and George Washington, Galileo Galilei ultimately made our democracy possible.

THEREFORE IN MAY 1993, I expected that what Saint Peter's Basilica would have been to my mother on our trip to Italy, the Florence museum room now containing Galileo's telescopes would be to me. As it turned out, however, I was lucky to get in to see the collection at all. A couple of days before I arrived, mafiosi had bombed the Uffizi, killing six people. In response, the entire city had gone on strike. When I alighted from the train, everything was still closed. Not sure what to do until my train left for Paris the next evening, I wandered over to the Basilica of Santa Croce—churches always stay open, of course—and spent some time admiring Galileo's magnificent tomb, the tomb they'd built him about a hundred years after his death, when people had come to realize he had been right.

The next day, a few hours before I had to leave Florence, the part of the Uffizi that held the history of science collection opened. The docent handed me an English-language self-guided tour brochure, and I moved slowly from item to item, pausing especially to appreciate the evolution the telescope had enjoyed in Galileo's own hands. Eventually I came upon a strange object—a relic, like the religious relics my mother and I had just visited all over Italy, perched on an alabaster pillar, protected under a beautiful dome of glass. This, the guidebook explained, was Galileo's middle finger.

It seems that when Galileo's body was moved, a century after his death, from a too ordinary grave (the grave of a heretic) to the grand tomb in the basilica (the grave of a hero), a devotee chopped off Galileo's middle finger and arranged this little shrine. A fellow named Tommaso Perelli had provided a Latin inscription for the marble: "This is the finger, belonging to the illustrious hand that ran through the skies, pointing at the immense spaces, and singling out new stars, offering to the senses a marvelous apparatus of crafted glass, and with wise daring they could reach where neither Enceladus nor Tiphaeus ever reached." (In Greek mythology, Enceladus and Tiphaeus, aka Typhoeus or Typhon, were giants who stormed heaven and led a revolt against the Olympian gods, only to be thunderbolted and crushed under Mount Etna by Zeus.)

Now, I knew that in Italy sticking your middle finger up doesn't mean what it means in the United States. But the more I thought about it—about Galileo's contentious nature, his belief in the righteousness of science, his ego, his burning knowledge that he and Copernicus were *right*, and especially about what the Church had put him through—the more amusing the middle finger thrust skyward seemed. I mean, of all the remnants, how perfect is it that with his remaining relic, the old man is eternally flipping the universe the bird?

Eventually I couldn't stand it anymore. I just burst out laughing, dropping the tour brochure on the floor. I picked it up and found the docent giving me a rather severe look. But I couldn't help myself. I started laughing uncontrollably again.

Somewhere on the crazy journey of the last few years, I stopped laughing at the image of Galileo's mummified middle finger and started thinking of it as a personal talisman. I would contemplate it to remind myself of certain propositions: That the mythical Galileo, a perfect man who could see beyond his own needs and his own psychology, never really lived—that uncomplicated heroes don't exist among the living. That all of us are struggling with the question of who we are. That sometimes people put you under house arrest because they honestly believe it is for the greater good. That it can be very hard in a moment of heated debate to tell who is right—it can take a hundred years and a thousand people to sort it out. As one person trying to get it right, sometimes the best you can do—the *most* you can do—is point to the sky, turn to the guy next to you, and ask, "Are you seeing what I'm seeing?"

CHAPTER 1

FUNNY LOOKING

YOU KNOW YOU'VE HIT UPON an interesting research topic when in a single week you get interview requests from both *Penthouse* magazine and Christian Life Radio. And you know you're doing something promising when both interviewers tell you they agree with your political stance.

That was my life, and this was my stance: Children born with genitals that look funny but work fine should not be surgically altered just because their genital appearance upsets or worries some adult. Big clitorises shouldn't be shortened, and baby boys with very small penises shouldn't be sex-changed just because their phalluses induce Freudian crises of confidence in their caregivers. People have the right to grow up with their genitals intact unless there is some dire medical emergency.

Not too surprisingly, the interviewers from *Penthouse* and from Christian Life Radio didn't agree with each other on the reasons for backing me up. The *Penthouse* guy's attitude was that no one should mess with your sexuality without your consent. He understood that the elective pediatric "normalizing" surgeries in question carry a lot of risk to sexual function and health and that sexual function and health really matter to one's life. Meanwhile, the Christian radio guy's position was that God doesn't make mistakes when he sends babies into the world. He didn't see the children in question as tragic failures meant to be corrected by some surgeon who apparently thinks he's more skilled or more compassionate than God.

I liked both their attitudes because, heck, it was 1998, and somebody in the real world was agreeing with me—just what the pediatric surgeons were telling me would never happen. *Two* somebodies, even! I mean, sure, the surgeons expected my message to be championed by my fellow academic feminists with hair as short and glasses as rectangular as mine, but that, they assured me, would be the limit of the "tolerant" population as far as genital anomalies were concerned.

"Honey," a very high-up surgeon once said to me right around this time, after hearing one of my talks, "you just don't understand. The parents in our clinics can't handle this." When a baby is born with ambiguous genitalia, he explained to me, the mother cries, and the father gets drunk. If you let a child with ambiguous genitalia grow up without surgery, he went on, the kid will commit suicide at puberty. That's just the way it is. *You can't change society.*

The surgeon telling me this apparently thought that I, a lowly assistant professor not long out of grad school, would simply yield to his authority, as did his medical students and residents and the parents of his patients. But I knew these surgeons' central argument—*you can't change society*—to be simply counterfactual. Here I was, a woman, in pants, with a PhD; my identity had come into existence only because society had been changed—and quite recently, too. But if I tried to tell these surgeons that together we *could* change society, they would see me as wanting to sacrifice babies on the altar of a radical social agenda. So instead I just took to answering these docs with a joking paraphrase of another uppity historian, Winston Churchill: History will vindicate me, and I will write the history.

I kept reminding myself, you can't really blame these guys for their myopia. Surgeons are taught to think in terms of days and weeks. They are trained to focus intensely on the surgical field immediately before them. I had an advantage here, being trained in history. Temporal depth perception spanning millennia means you not only can see the potential to enact meaningful social change in terms of identity politics, but also you can see that, no matter what you do, you'll eventually be forgotten. There's something really liberating in knowing you don't matter.

HUMAN SEX COMES in two big themes—male and female—but nature seems to enjoy composing variations on those themes. Some sex variations occur at the level of sex chromosomes, some at the level of hormones, some at the level of hard-to-detect internal structures, and some at the level of anatomical parts you can see with the naked eye (assuming your eye isn't the only thing that's naked). If you call all of these variations intersex, you can then ask how common intersex is. That's a question people love to ask. The problem is that to answer that question, one has to first decide how subtle a variation to count. How small should a penis be to count as intersex rather than male? How big a clitoris should count? How subtle a difference in hormone receptors? The truth is that human sex isn't simple. Human sex is practically fractal.

Nevertheless, wherever nature draws unclear boundaries, humans are happy to curate. And the specialist curators of sex tell us this: In America today, about one in two thousand babies is born with genitals so notably intersex that a specialist team is immediately called in. About one in three hundred babies has genitals unusual enough that the average pediatrician will give the parents a referral to a specialist. If you add up all of the dozens of kinds of sex anomalies—including incredibly subtle things you might never know you had without the benefit of a lot of fancy medical scans your insurance company probably doesn't want to cover—the frequency of intersex in the human population comes to about one in a hundred.

I was twenty-five and in graduate school before I ever learned that sex anomalies happen in humans. Years later, when I had become an assistant professor, I found out that a boy I'd grown up with had been born with ambiguous genitalia. My mother revealed this to me after reading some of my work, and when she revealed it, I asked her where that old friend was now. I told her I wanted to connect him with the support groups that had sprung up.

"Oh no," she answered, "we never told him."

This was exactly what I was then documenting in my work—how babies born with sex anomalies were "fixed" and then promptly swaddled in shame, secrecy, and lies.

The topic of hermaphroditism had been suggested to me early in my graduate career by Fred Churchill, the man who became one of my dissertation directors. Fred didn't understand feminism—I remember that, one day, he admitted, befuddled, "I still don't get what's *feminine* about *feminism*"—but he believed in supporting students' interests, and mine was gender. Following Fred's suggestion, I thought I should look at the history of embryology and Charles Darwin's study of barnacles. (Those barnacles were hermaphroditic.) But Fred kept saying, "Look at medicine." *Why?* I wondered. *Hermaphroditism doesn't happen in humans past the embryonic stage.* Then a second graduate advisor who knew I was on to hermaphrodites also said, "Look at medicine."

So I looked. I went over to the university's main library and pulled the second *H* volume of the worldwide index of nineteenth-century medical literature, *The Index-Catalogue of the Library of the Surgeon-General's Office*—the comprehensive government-run index of medical reports that eventually came to be known as PubMed. To my shock, there I found pages and pages listing case reports of human hermaphrodites from the 1800s. Hermaphroditism in newborns, in children, in adults, and in dead old people who just happened to get dissected by some lucky anatomist. Cases recorded in England, France, Germany, Poland, the United States, India, and just about every other locale known to Western medicine.

Why had I never heard of this?

I decided then to do my dissertation work on the history in science and medicine of what happened to the people labeled hermaphrodites in late-nineteenth-century Britain and France. These were people who either had mixed-sex anatomy externally or who were found via surgery or autopsy to have one apparent sex on the outside of the body and the other on the inside. There was a nice set of about three hundred papers, mostly medical journal articles, that I could use as my primary sources, and the period was particularly interesting because it was a time when homosexuals and women had started agitating to loosen gender norms. I figured it would be interesting to see what medical and scientific men, who tended to be politically conservative, did when confronted with *natural* sex blurring, given that their abhorrence of women's suffrage and men-loving men was based on the "fact" that such things were fundamentally unnatural. Britain and France also made practical sense because they would be nice places to go on dissertation grants, and I could read English and French. Truthfully, the French was sometimes hard, but Fred helpfully suggested that translations often go faster if you drink a bit of wine from the region where your texts originated. (I still can't drink French wine without thinking about sex.)

During a cold snap in February 1994, I flew to Bethesda, Maryland, to the National Library of Medicine, to collect more material for my work. On the plane, I was seated next to an army doctor who was in charge of HIV management for the military. He asked me what I was working on, and upon hearing me describe my project, he sternly said to me over his reading glasses, "Hermaphrodites are not a marketable skill." Even so, it was obvious he wanted to hear more.

In history as in science, you look for what changes and what is stable, you look for correlations, and you pray to find evidence of causation. I had a lot of data to work with—more than I could manage with just my brain and marginalia—so I took my three hundred primary sources and created a computer database of their components to see what patterns I could find. I tracked specific patients based on the demographics the medical reporters gave about them, to see if they might pop up more than once in the medical literature. (This enabled me to "watch" one poor nineteenth-century Frenchwoman with a herniated testicle wander from doctor to doctor looking for some productive help.) I cataloged individual scientists and doctors to track who was reporting what when. And then I cataloged each individual article's specific report, opinion, or theory—if it expressed one—about Adam's apples, beards, behaviors, breasts, clitorises, erotic orientations, gonads, hair, interests, labia, menstrual signs, penises, prostates, scrotums, skeletal proportions, uteruses, voices, vaginas, and vulvas. Doing this allowed me to sort, for example, by breasts, to see whether there were patterns or shifts in what medical and scientific men thought about the meaning of feminine breasts in a supposed man, or the meaning of perky nipples in a manly woman.

I ended up with notebooks filled with hundreds of dot-matrix-printed, three-hole-punched pages of data—and a growing clarity about what had happened. Over the course of the nineteenth century, there were more and more doctors, more and more people were going to doctors, and the medical profession was enabling and rewarding publication, so the incidence of reports of humans with sex anomalies shot up and up. This alarmed the medical men even as they were tremendously fascinated. After all, social order was based on the presumption that it reflected natural order, yet nature seemed to have a very bad habit of not picking up after herself.

Upon expert examination, a surprising number of people showed obvious external ambiguity in their genitals. Others mixed a masculine chest and beard with feminine genitalia or feminine breasts with a penis. Still others looked all well and good on the outside but were discovered (during surgery or autopsy) to in fact have the other sex's organs inside. A very few even claimed they were clearly female as children but had become quite manly at puberty.

Notably, most of the people who showed up in the nineteenth-century medical literature with sex anomalies seemed either not to know or not to be terribly bothered. Sometimes a woman wanted an explanation as to why her vagina seemed to be too small to accommodate her husband's penis. A man and his wife wanted to know why the husband seemed to bleed on a monthly cycle, while a chambermaid in another city wanted to know why she never seemed to menstruate. On rare occasions, a child or an adult was reassigned from one gender role to another after being diagnosed as a victim of "mistaken sex." (The "mistake" was often blamed on some stupid midwife.) But most often, nineteenth-century patients with such traits just went on with their lives as before their clinical exam, probably never knowing just how high they'd made the doctors' eyebrows rise.

Late-nineteenth-century medical and scientific men had little interest in changing social mores just because nature was turning out to be a bit churlish where sex was concerned. Instead they opted to impose order on nature as best they could. They came to a consensus by the 1890s that made sex look a lot more organized than it really was. They decided that "true sex" would depend only on the gonads: If you had ovaries, no matter how many manly traits they came with, you were really a female (though perhaps a female pseudohermaphrodite); if you had testes, then you were really a male (though perhaps a male pseudohermaphrodite); and only if you were absolutely positively proven to have both ovarian and testicular tissue were you a "true hermaphrodite."

Conveniently, because biopsies were not possible at this time, the only people who could be conclusively labeled true hermaphrodites were at least castrated and more often dead. This approach pretty much neutered any threat that hermaphroditism might have posed to the social order. Virtually everybody was *really* male or female; sex just *looked* blurry to the untrained eye. Problem solved. As I documented this cleverly constructed solution to blurry sex, I decided to label this period the Age of Gonads.

Of course, not too much later—in the 1910s—biopsies *did* become possible, and the Age of Gonads had to cease. Suddenly doctors could conclusively diagnose working ovaries in men, working testes in women, and ovotestes in both—not a happy thing unless you're a gender radical. So again doctors did what they had to do to preserve the two-sex social order. Although they still categorized a patient's "true sex" according to gonadal tissue, in practice they classified patients according to which gender was most believable. If an attractive housewife happened to have testicles, no one besides her doctor needed to know her diagnosis of male pseudohermaphroditsm. If a man really was menstruating, you just quietly took his ovaries out and hoped no one found out about his insides. Doctors continued to clean up nature's little indiscretions and thus take care not only of individual bodies, but also the social body.

Given the way intersex could always threaten a sexist two-gender society, this approach of "cleaning up" nature's sexual "mistakes" persisted in American medicine, leading in the 1950s to the key collaboration at Johns Hopkins University of Lawson Wilkins, the founder of pediatric endocrinology, and the psychologist John Money. Together Wilkins and Money formalized this cleanup approach to sex anomalies. Modern medicine now sought to reinforce the "optimum gender of rearing" by early management of children born with sex anomalies by means of "sex-normalizing" surgeries, hormone treatments, delicate euphemisms, and sometimes lies.

This, then, was the system of concealment that led one day to my mother's remark to me about my childhood friend: "Oh no, we never told him." This was also the system that led to a lot of really angry intersex adults who discovered that they had been harmed by the medical care meant to "save" them and who knew the same basic system was still being used on children who would likely grow up as hurt and angry as they were. In the early 1990s, a core group of these people formed the intersex rights movement I eventually joined. Some of these intersex adults had been physically harmed—left with damaged sexual sensation, incontinence, or repetitive infections. Many had been psychologically harmed—left with a sense of having been too monstrous for their parents to accept as they came, of being sexually freakish, of being fountains of familial shame. All were left with a burning desire to try to save others from going through what they had.

I knew about the nascent intersex rights movement as I worked on my dissertation in Bloomington, but I mostly ignored it. My graduate school professors had taught me that history is about the dead. Yet shortly after I started to make public my research findings, I found myself in dialogue with Bo Laurent, the founder of the Intersex Society of North America (ISNA). As I recall, I first talked to Bo by e-mail using her public activist name, Cheryl Chase, a name she had hastily chosen as a nom de guerre early in her activism. As soon as I learned that another intersex activist named Bo Laurent lived in the same region of northern California as Cheryl Chase, I tried to bring Bo and Cheryl together. She politely explained I was trying to introduce her to herself. I easily fell into the habit of publicly referring to my work colleague, the leader of ISNA, as Cheryl, and privately calling my new friend Bo.

Bo was fascinated by my findings and immediately proved herself to be a voracious and intelligent audience of intersex history. Besides being geeks and history buffs, Bo and I shared what I later realized was a rare trait in activists: a belief in evidence even when it challenged our political goals. While it would have been convenient for Bo if I could have told her that no one in the nineteenth century sought surgeries to make their own genitals look or act more normally, she readily accepted my data that a few people did. She also readily accepted my conclusion that we couldn't know much about why a few people in the 1800s seemed distressed about their own anomalous sex while most seemed fairly unconcerned. The data was incomplete, and rarely could it tell us about something as obscure as the psychological history of an unnamed, one-time patient who probably died before we were born.

I realized as soon as she told me her personal history that in Bo I had met someone who actually represented in the flesh that old Victorian compromise with nature that I had documented in my dissertation. That is to say, because of the fears of my dear Victorian doctors, Bo was to be counted as a true hermaphrodite simply because she had been born with ovotestes. (An ovotestis is usually kind of like an ovary wrapped up in some testicular tissue. And no, even if you have one, you can't make both viable eggs and sperm.) Even though genetics, endocrinology, and other sciences had traversed a thousand miles from the Age of Gonads to the age of ISNA, that Paleolithic terminology based on gonadal tissue was still in place. (Bo and I later successfully worked to get rid of it once and for all.)

Bo had also been born with ambiguous genitalia. The doctors were so upset when that baby had come out, in 1956, that they kept sedating the mother while they figured out what to do. After three days, they finally let Mrs. Sullivan wake up and go home with the baby the Sullivans decided to call Brian—a boy understood to be healthy but sexually defective. They also sent Mrs. Sullivan home with strict instructions not to let anyone know that this child had a malformed phallus and a vagina. Mrs. Sullivan understood that she had to do all the diaper changes herself, and hide the disaster forever.

When Brian Sullivan was about a year and a half old, a team of doctors at New York–Presbyterian Hospital discovered via exploratory surgery that Brian had a uterus and ovotestes. The doctors realized that these organs might mean that they could turn this genitally unconvincing boy into a fertile girl, so they completely removed the phallus of the child, who at that point, on their strong recommendation, became Bonnie Sullivan. When she was eight, without anyone explaining anything to the bright and perceptive patient, Bonnie Sullivan was sent back to the hospital and her testicular tissue was removed. (This later surgery, removing the testicular portion of the ovotestes, may well have prevented Bonnie from getting a form of testicular cancer. We have never argued against that kind of surgery, which evidence really did show was lifesaving.)

When she became a sexually active lesbian as a teen, Bonnie Sullivan quickly figured out that she was missing a clitoris and, more important, missing sexual function. She could not have an orgasm. To make a harrowing long story short, eventually she sought out people like herself, found them, and marshaled her lesbian feminist political consciousness to fight the system that had taken so much from her. In 1995, as she broke with a past that had never really been under her command, Bonnie Sullivan took on a new name. Bo was a short form of Bonnie. She chose her last name, Laurent, as a tribute to the deaf activist Laurent Clerc and in remembrance of her own grandparents, who had been deaf. Bo understood that intersex had to go where deafness had gone: from the realm of the "defective" to the realm of rights. She started looking for peers and allies anywhere she could find them.

Not long after we started communicating, Bo asked me—as she asked perfect strangers all the time—to help her change the medical system for treating intersex. At first I resisted. As I explained, I was just a historian, somebody who deals with the past, and a newly minted one at that. I also thought it would be a little odd for me to get involved in a movement for rights that did not directly affect me—although having been raised with a brother who is identified as black in an otherwise white family, I did get that civil-rights movements need anyone with the power to help. Hearing Bo's gentle plea, I also found myself remembering something my mother had said to me before I left for Indiana to study the history and philosophy of science: "I hope you study philosophy, because maybe then you can at least help people think more clearly. History is just about what's already finished." The implication: there's no social good in being a historian.

In any event, Bo pushed me, in her typically blunt yet soft-spoken way, to consider putting my feminist money where my mouth was. Just look, she said, at what today's medical books say. That was easy enough. Right around the time I had been told hermaphrodites are not a marketable skill, I had fallen in love with a medical student in Bloomington. I had moved in with Aron a few months later, when he had to start his clinical rotations in Indianapolis. As a consequence, my history books were on one side of our little apartment's living room, and Aron's medical books were on the other, so, on a break one day, I pulled his books and looked up the standard of care for genital anomalies. This would have been in 1995.

Basically, what Aron and his classmates were being taught, in the latest medical books, was this: If a baby is born with a large clitoris, she might turn out to be a lesbian, so you have to cut down her clitoris. If a boy is born with hypospadias—wherein the opening of his urethra is not at the tip of the penis but on the underside or down near the scrotum—he will not be able to write his name in the snow next to other little boys, and then he might turn out gay. Therefore you have to do a "corrective" surgery to make sure he can pee standing up. Mind you, this surgery failed so often that doctors had a special term for the men in whom it failed. They were called hypospadias cripples, because life is tough with a surgically scarred, infection-prone penis, but, the urologists insisted, you had to *try* to get that boy to pee standing up. *Or else.*

Huh?

When Aron got home from the hospital that day, I showed him the relevant passages in his books. I then asked him if he had written his name in the snow next to other little boys. "No," he said, putting both his hands to his cheeks in a mockery of the book. "Do you think I might be gay?"

We were both pretty stunned. *OK*, I thought, *I'll help Bo.* I'll go to these doctors with her, and we'll reason with them. We'll use data from history and medicine. We'll say, Look, this system was developed before gay and lesbian rights, before an appreciation for women's sexual health, before modern medical ethics. There's no evidence to support this heterosexist system of shame and secrecy as good for the patients, and there is evidence, albeit anecdotal, that this approach is unnecessary, unsafe, ineffective, and deeply harmful. Parents can learn to raise these children as boys and girls without cosmetic genital surgeries; we knew of some who had. Intersex children can be told the truth about their bodily differences, and they can grow up to decide for themselves what they want to do with their own bodies in terms of optional medical care.

And to all this the doctors would say, "Oh, gosh, yes, we've been meaning to fix this. Thanks!" And it would all change.

Aron kept shaking his head at me. "Medicine is not like that," he'd say. "Medicine is often not rational and not evidence-based." But he'd say it not so often nor so firmly as to discourage me. Aron has a theory that, for the sake of progressive change, people should sometimes be left in a state of productive naïveté, and that was his prescription for me in this case. So I plunged ahead, thinking reform of the system would take about six months, maybe twelve at the outside.

Bo must have figured out early that she needed to keep me motivated, so she did something really smart. She brought me around to meet as many intersex people as she could and made me listen to their stories. I started listening with a tape recorder and often found myself losing it, breaking down crying. In the medical literature, all of these people were "lost to follow-up." They had dropped out of treatment and were invisible to medical surveillance, so what few outcomes studies there were (and there were almost none into adulthood) did not count them. No one really understood what these people had been through. I mean, a few people had heard these individuals' stories one-on-one, but no one except maybe Bo and a couple of other intersex people had heard how this population had suffered collectively.

There was Martha Coventry, who had had most of her "enlarged clitoris" cut off when she was six, because her parents were worried about gender confusion. There was David Cameron Strachan, who as an adult had been diagnosed with Klinefelter's syndrome (XXY chromosomes) and had been shot full of testosterone by a doctor who maybe had thought that upping his sex drive would turn David from gay to straight. (Instead, it had made David uncomfortably hairy and horny, and he'd headed right to the San Francisco Castro District, just as HIV was arriving there, too.) There was a woman whom I'll call Beth Lawrence, who had learned that she'd been born with testes when she had opened a sealed envelope that a doctor had told her not to open but to give straight to her next doctor. Beth had opened it in the parking lot outside her doctor's office and found a medical journal article about her, her sisters, and her cousins, none of whom had been told they had testes. The article featured a large photograph of Beth as a teenager, standing naked in a medical setting, with her head cropped off—I suppose to keep her from being recognizable as Beth Lawrence.

Of course, the pose and the crop also kept her from being recognizable as fully human. Beth still had that journal article, which she kept in the original envelope. Sitting in her backyard with her and Bo and looking at the article, I realized something bizarre: When we'd first met, I had insanely expected that Bo would walk up to me naked in black and white with her eyes blacked out. That's how the medical journals had led me to think of contemporary intersex people. Even though I had been conversing with Bo by e-mail, I had subconsciously expected her to show up naked with a helpful grid behind her, her arms and legs splayed so as to show off her proportions to the medical student gazing upon her in the book.

The fact that intersex adults actually meet you in full color, with their clothes on, with families and lives and mortgage debt and a lot of pain and trails of therapists who didn't even believe their unbelievable histories—this all just made me more and more upset. Pediatric endocrinologists and pediatric urologists came to my invited academic lectures and told me they felt sure *their* patients were all doing fine. That just made me disoriented. Where are these happy patients? I asked. They told me they could not say. Privacy prevented it, they said. Ask your happy patients if they will talk to me confidentially, I said. Oh no, they said. We wouldn't want you upsetting them. "But," one assured me, "I know my patients are doing great because I get invited to their weddings."

"That's sweet," I answered, trying to hide my sarcasm. "How many have you been invited to?"

"Um . . ." He hesitated. "One, I guess."

Bo also introduced me to a few people she'd found who had grown up with ambiguous genitalia, having gone "uncorrected" for various reasons. There was one woman whose sexually sensible mother had decided there just wasn't any good reason to follow the doctor's advice to shorten her young daughter's noticeably long clitoris. There was another who had been too sickly as a child to be sent into elective surgery and so had escaped, growing up unharmed and ultimately pursuing a career in medicine. A third I met because she came to hear a talk I was giving in her town. This woman came up to me after my lecture to say she thought she might be intersex, because her clitoris was bigger than most. "How big?" I asked. She showed me her little finger and said she used her clit to penetrate her partner's vagina, quietly adding that they both enjoyed it.

Whenever I felt my energy starting to flag, the universe seemed to send me something to renew my sense of urgency. One day Aron found himself having to calm down a young woman who had just been told by one of Aron's internal medicine residents that she was really a male pseudohermaphrodite—that she was really a *man*—because he had figured out that she had testes inside. Aron called to get from me the name and number of a representative of the Androgen Insensitivity Syndrome Support Group. Another day, I got a call from a nineteen-year-old man who had just found out he had ovaries and a uterus inside of him. One of his doctors was suggesting he get a "sex change" so that he could be a woman and have a baby. I knew just how badly that doc wanted that publication.

"Do you want to be a woman? Do you think of yourself as a woman?" I asked the young man.

"No," he said, "but the doctor says because I have ovaries . . ."

"Look," I told him, "I don't let my ovaries tell *me* what to do. I don't think you should let your ovaries tell *you* what to do."

I especially remember sitting at work in my Michigan State University assistant professor office one day and out of the blue getting a call from a weeping pediatric nurse I had never met. She calmed down just enough to explain to me that they had a baby in their pediatric ICU who had been sent into surgery to make her genitals look more normal. The baby had gone into surgery healthy. The anesthesia had gone wrong, as it sometimes does in babies, and now she was going to die. This little girl was going to die just because her clitoris had been "too big." And now her parents would have to live with that twisted memory of guilt, shame, and grief. I knew that story would never enter the medical literature. Surgeons rarely report when it all goes wrong. They have their own guilt, shame, and grief, typically left as unprocessed as the parents'.

DAY BY DAY, the Intersex Society became increasingly intertwined with my existence. When Bo decided to legally incorporate, she asked me and Aron if she could use our home address, because we had a stable residence and were actively involved. Upon incorporation, I became the first chair of the board of directors, and if I remember correctly, Aron became vice president. We didn't expect him to actually do anything; we just needed a certain number of signatures, and we thought that having MDs on the board would help persuade the IRS to give us nonprofit status. Bo's partner, Robin, took some other executive title. I started joking that our first task should be to work on the board member–to–bed ratio. The final addition to the board was another Michigan State medical faculty member, Bruce Wilson, one of the first pediatric endocrinologists to say we were right.

We worked with a small army of other intersex activists who were also out there pushing for change. Many of them, like Max Beck, Mani Mitchell, Emi Koyama, Hida Viloria, and Tiger Devore, told their own stories on television and in documentaries and spoke to any group that invited them. Early on, Bo gave me a handful of "phall-o-meters" to start handing out, a little tool developed by the intersex activist Kiira Triea. The phall-o-meter showed graphically how doctors decided whose phallus would be cut and to what length to make them fit social norms. It was a to-scale measuring stick that went from "just a girl" (for a small clit) to "fix it quick" (in between) to "phew, just squeaks by" (a barely acceptable penis) to "OK" to "Texan" to "Wow, surgeon!" (for the big 'uns). I handed these out on all sides and left them behind everywhere, between the pages of in-flight magazines, in the stalls of women's bathrooms on campuses I was visiting, and in the hands of all the surgeons I ran into. The male surgeons just loved them.

Bo had taught me this blitzkrieg method. We simply took every opportunity that came along and sought out any others we thought might work. She encouraged me to keep working the academic angle, and I did, doing scholarship in support of the movement. The last chapter of the book based on my dissertation provided an extensive ethical critique of the modern-day management of intersex. That it had Harvard University Press's name attached definitely helped. I spun off that last chapter as an article for the *Hastings Center Report*, the journal of the leading independent medical ethics institute. The next book I published was an edited anthology called *Intersex in the Age of Ethics*.

For that collection, Bo and I wanted a front cover that showed the contrast between the monstrous medical image of intersex and the real lives of intersex people—to make the point that you never know who around you is intersex and the point that the medical approach is what makes someone a monster. We had realized how powerful images were in getting people to change their thinking. So we took photos that all the contributors—intersex and non-intersex—gave us of themselves, and put those, all mixed up, on the front cover. A few were bare-chested men; most were fully clothed. You couldn't tell who was intersex and who wasn't. For the center of the montage, we wanted a classic medical image—naked, eyes blacked out, against the grid—but I didn't dare use a real image and reexploit someone. I can't remember if Bo or the publisher suggested it, but one of them said to me, "Why don't we do a picture of you, Alice?"

So I paid a university photographer whom I'd come to know fifty bucks to meet me at his apartment and photograph me naked standing in the "medical pose" with a band of paper meant to look like a hospital ID bracelet taped around my right wrist. He then used Photoshop to put a grid behind me and a black band over my eyes. He also blurred out my naughty bits. (I didn't have tenure.) When my friends and students saw the book, they immediately recognized me. So much for the idea that the black band makes any difference! I just told them I do nudity only if the plot requires it.

The plot required so much. Time, money, and lots of personal effort to keep the activists from infighting due to jealousy, philosophical differences, and pent-up fury. And so much effort to keep Bo from falling into another black abyss of posttraumatic depression. Because I could write and speak well, I did one television show after another, quickly learning what to wear (no white and no small prints; lots of powder and bright lipstick; a serious look with a kind smile) and how to wrap a clear message around a killer story. I wrote newsletter material, teaching materials, and fund-raising appeals. I learned how to ask people, point-blank, for money to support us. Money was always short; Aron and I regularly dumped in infusions of cash, trying to keep enough in the till to keep Bo from having to do other work, so she could stay focused on ISNA. A sizable percentage of the donor list was made up of our personal friends and family members. Bo spent down her life's savings as we pressed on.

Now and again, we caught a break. Someone would invite us to speak at a place where there was a doc with enough doubts that she or he would then sign on to help us. Someone with power would have an adult child who was gay or lesbian, enabling that powerful person to appreciate at the gut level the way that discrimination against sexual minorities manifests in every bit of life.

A big break came in 2000 when John Colapinto published the "John/Joan" story in his blockbuster book *As Nature Made Him: The Boy Who Was Raised as a Girl.* Colapinto's work brought to national attention the story of one male child whom John Money had recommended sex-changing after the baby's penis had been accidentally burned off during a medical circumcision at eight months of age. The patient, now known as David Reimer, had not been born intersex, as most of Money's patients had been; David Reimer had been born a typical male, with an identical twin brother. But after his circumcision accident, the family was referred to Johns Hopkins and, on Money's recommendation, the baby had been surgically and socially turned into a girl named Brenda. After all, a boy without a penis (or with a very small one) couldn't grow up to be a real man! At least that's what Money et al. had been saying for years. Money must have been thrilled when he encountered the Reimers: Here, in a set of identical non-intersex twin baby boys, was the perfect case to prove his theory that gender identity development depended primarily on genital appearance and upbringing. If one of the Reimers' twin boys could be turned into a girl, this would be the Hope Diamond in Money's crown.

Thanks to Money's desire to use David Reimer to prove that gender is mostly a product of genital appearance and nurture, not inborn nature, Reimer had gotten caught in the Johns Hopkins intersex vortex and had had the same history of shame, secrecy, loss of function, trauma, and anger as many intersex adults. Importantly, Reimer also failed to prove Money's theory. As little Brenda, he kept acting boyish, and upon being told the truth of his medical history as a teenager, he immediately declared himself a boy and socially became a boy again. Nevertheless, Money simply lied about the outcome, leading everyone to continue believing his experiment with "Brenda" had worked.

Although *As Nature Made Him* entailed great coverage of our work at ISNA, Colapinto's account moved people for a reason we had come to resent: The public was ever so upset that a "real" little boy had been turned into a girl. They were upset about the sex-change of a non-intersex child and about having been led to believe that gender is a product of nurture, not nature. To us, the primary issue in these cases wasn't the nature of gender. Yes, the reason all these kids—Reimer and his born-intersex cohort—had been traumatized was because of a wrong theory of gender that said that we can make you into a boy or a girl if we just make your body look convincing in infancy. But the trauma for most of these folks didn't come from getting the wrong gender label as a baby.

Bo and I knew what the clinicians knew—that most intersex people kept the gender assignments they were given, whether surgeons made their genitals look typical for their gender or not. And we knew that people who changed their gender labels as teenagers and adults did not find misidentified gender to be the core of their suffering. The problem in intersex care wasn't a problem of gender identity per se. The problem was that, in the service of strict gender norms, *people were being cut up, lied to, and made to feel profoundly ashamed of themselves.* Bo said it as plainly as she could: Intersex is not primarily about gender identity; it is about shame, secrecy, and trauma. Doctors were so obsessed with "getting the gender right" that they didn't see that they were causing so much harm. If they could have obsessed less about gender identity outcomes in these cases and focused on actual physical and psychological *health,* they might have done a lot less damage. They needed to stop treating these cases as gender identity experiments and start treating them as *patients.*

But most people didn't want to hear about shame, secrecy, and trauma when we talked about intersex. They wanted to hear about the nature-nurture debate. Just like John Money, they wanted to use intersex people in the service of their theory building about gender identity. All that happened when people started to take the nature of gender identity seriously was that docs stopped turning boys with micropenis into girls and started pumping them full of risky drugs to try to get their penises to grow bigger. The clitoral surgeries—those kept up.

It would be easy to fall into the belief that these were all evil doctors. Truth is, they were basically good people. They had been told in their medical training the same story the surgeon told me early in my work: If you don't do this, these kids will kill themselves at puberty. Based on this mythology, they believed they had to do early genital surgeries. Bill Reiner, a urologist who had trained at Money's gender clinic and who later turned against Money's approach, told me that he'd once tried to find evidence that kids had killed themselves as a result of being left "uncorrected." Like me, Bill couldn't find it.

The myth of teen intersex suicide was part of what my friend Howard Brody, a physician-ethicist, took to calling the maximin strategy in medicine. When a doc "maximins," she maximizes the number of interventions in the hope of minimizing the odds of the worst possible thing happening to that patient. You operate out of fear of the worst-case scenario. Howard had traced this in obstetrics, and had shown how obstetricians were actively harming mothers and babies during normal births in an effort to keep them from dying. They were throwing every possible intervention at them, because then, if the mother or baby died during a birth, at least the doctor had tried everything. It was just a natural coping strategy in a stressful situation. But when you looked at the aggregate *evidence*, the interventions meant to prevent the worst harm actually resulted in *more* net harm.

That's what was going on with these intersex specialists. They were afraid to "do nothing," as they put it. We said, "Don't do *nothing*; call in mental health professionals to help with shame, fear, and grief." But the doctors said they didn't know whom to call. And it was true; Money, a psychologist, had popularized this whole system of care in his writings, but it had really been founded and disseminated throughout the medical world by Lawson Wilkins, the founder of pediatric endocrinology. Instead of teams of psychologists to help intersex people and their parents, there were only pediatric endocrinologists, who knew little of psychology except what they had been told: Gender is all about genital appearance; call the surgeon.

When I would ask treating physicians, "What is the goal of pediatric intersex treatment?" I was amazed at how often they could not articulate an answer. It was clear that they were operating from a combination of institutional inertia and an impulsive (beneficent) need to quiet down parents they thought might get upset. It would have been much easier if all these doctors had been evil. Instead, they were good—human, scared. They tried hard to write us off as evil, but when they met us, they realized that we were also good—and human and scared.

OK, so I'm not so sure John Money was good. He had used and abused so many of my intersex friends who'd had the misfortune as children to end up in his Johns Hopkins clinic that we called the place the Death Star. Money had *known* that David Reimer's life had not turned out well, that he had never been a straightforward girl, and that as a teen he had reverted to being a boy. He had lied about and to Reimer and hurt many other people in the process. It was tempting to try to take Money down, to go after him personally.

But Bo was smart again. Even though the one time she'd met Money in person at some cocktail party, he'd started screaming at her at the top of his lungs, she decided that we would not engage in ad hominem attacks, not even against Money (except in private, over a lot of alcohol). She said if we take down an individual, the system has not changed. That person becomes a scapegoat, and nothing really changes. And she was right. Reporters would come to me and say, "Well, Dr. So-and-So says that he now knows John Money was wrong about gender, so now he agrees with you, and there are no more ethical issues." Meanwhile, Dr. So-and-So would be routinely performing surgery on baby girls with big clitorises and telling adolescent girls with testes that they had "twisted ovaries" that needed to come out, with no evidence for the supposed medical necessity or benefit of these approaches, especially when compared to the risk of harm.

But we were seeing signs that we were making progress. By the early 2000s, journalists started finding it impossible to locate a doctor who would say, on camera or in print, that we were wrong about anything. And they found more and more who were willing to say we were right. Articles and op-eds started appearing in medical journals calling for outcomes research to determine what had really happened. Medical students were rising up against being taught the old model; we heard of them handing their professors our literature and demanding that they be taught by someone whose ethics were in keeping with what they were being taught in their ethics classes. Our activist allies were being increasingly invited not just to local churches and synagogues to speak of their lives, but to medical centers, too. Little by little, Bo and I were being invited to give not just talks at medical events, but to deliver grand-rounds presentations at children's hospitals and keynotes at medical conferences.

IN MY OWN SCHOLARSHIP, I branched out from intersex in response to a question from Bo: How much of the reaction to babies born (as she was) with ambiguous genitalia is about fear of sex, and how much is about fear of abnormality? I decided to look at conjoined twins, thinking that by studying them I could control historically for sexual attitudes. Silly me! I soon found that conjoined twin babies, like intersex babies, had gotten tangled up in adult sexual phobias. As I researched the history, it became clear that conjoined twin separations, rather than being based on evidence of what would leave the twins best off, had often been based on an adult sexual fear: If you left conjoined twins to grow up conjoined, they might never have sex! Or they might even *have sex*! I remember bells going off when I ran across one news report of conjoined infant sisters from Guatemala; a UCLA surgeon told a reporter that when he made the final cut that separated them, he announced to his team in the OR, "We now have two weddings to go to." Hello. Happy weddings as a measure of whether the medical intervention was justified? That sounded very familiar.

Once I assembled the data about the history of medical responses to conjoined twinning, I was shocked to realize not only that sex phobias were sometimes driving separations, but also that in many cases separation likely left twins *worse* off, with more impairment and shorter life spans. Were they better off psychologically? Who could tell?—because, as with intersex, though surgery was often done for putative psychosocial reasons, no one was really looking at long-term psychosocial outcomes of those left alone or of those "fixed." Yet if we looked across a broad span of history at what was known about people left conjoined, it turned out that being conjoined was often probably better than being left with massive surgical damage (or, um, left dead). Conjoined twins old enough to give their own views said was that they were OK with their condition; they understood that it wasn't normal for other people, but it was normal for them. Only one set of conjoined twins in history, Ladan and Laleh Bijani, has ever elected separation for themselves, and even in that case, there is reason to believe the twenty-nine-year-old sisters may not have had an accurate understanding of the level of risk associated with separation of head-joined twins like them. Just after the sisters' deaths from surgery, the lead surgeon involved told reporters, "At least we helped them achieve their dream of separation."

About halfway through my study of the surgical treatment of conjoined twinning, I realized that, if I let the evidence lead me where it seemed to go, I was going to have to start arguing against some conjoined twin separations—not all, but ones that looked as though they weren't in the patients' best interest as far as the evidence went.

That's when I realized I'd better grow my hair out.

By then it had become clear that some of the resistance among the doctors we were arguing with over intersex was their perception that Bo and I were really just champions of the "gay agenda." We were really just there to recruit their infant patients, for the toasters I hear you get when you convert a certain number of people to being gay. We were read as queer. Hell, Bo *was* queer, and clear about it. (I was often presumed to be her romantic partner.) So our intersex "agenda" was being read by many doctors as really being about lesbian, gay, bisexual, and transgender (LGBT) rights. To be fair, their reading was not without cause. Intersex had quickly gotten wrapped up in the LGBT rainbow. Many early intersex activists identified as gay or lesbian—or simply queer—and their political consciousness about LGBT rights had caused them to be politically astute about intersex, too. Non-intersex LGBT activists had also helped the intersex-rights movement from the start, because they immediately understood this to be an issue of discrimination against a sexual minority. And homophobia was very clearly motivating a lot of the old clinical regime. How else could you explain outcomes studies that measured *not* whether women could have orgasms after clitorectomy, but whether these women were getting penetrated by men?

Still, it was highly unlikely that we could undo homophobia in a short time, so how were we going to get around the clinicians' resistance? It became clear that it might help if we tried as hard as we could to take the perceived gay agenda off the table. That meant I had to stop being read and easily written off as a lesbian feminist. If I was going to argue for something as radical as letting girls keep their big clits and sometimes letting conjoined babies live until they died naturally, I was going to have to look less socially radical and try to act less aggressive—less "manly." So I grew out my hair and invested in some pretty dresses and even pantyhose and pumps. I started categorizing surgeons into two classes: those powerful enough to be worth shaving my legs for and those not. I started carrying around an index-card reminder to myself: "Talk slower. Don't shriek." To my mother's delight, I even started wearing lipstick off camera. When one of my old friends discovered me in this drag, I confessed that, yes, I had, in fact, become a whore for social justice.

And it helped. It also helped that I started cracking a joke at the start of every medical talk: "I'm not a doctor, but I sleep with one." It helped that we started talking with doctors about the very real stress they were feeling. It helped that we started praising them effusively for every baby step forward. It helped that we introduced one reformer to another, so that they had some peer support in their little revolution. It helped that we made them feel special, invaluable, and liked. We started paying attention to relationships, having meals with the people we were trying to change, or at least coffee. It helped that we started treating them as humans.

And it really helped that—unlike most of our putative academic political allies, who wanted to just spew cute slogans and academic postmodernist horseshit—Bo and I mastered all the scientific and medical evidence and language we could. We learned enough biochemistry and anatomy to keep up with every question or argument thrown at us. We asked clinical researchers for data in advance of their publications so as to sound one step ahead of the curve. When doctors plagiarized from my or Bo's work, rather than fighting for our citation, we shut up and smiled and let them believe they had come to it on their own. We pushed as many people as we could into the limelight and stayed back more and more, to make our ranks look as big as possible. With Bo's expertise in computers, my writing skills, and our joint ally building, we looked very big.

At some point, Bo and I had the discussion about whether, if the evidence showed people were better off with cosmetic genital surgeries done in infancy, we would accept it. We came to the same conclusion: If most of the women who'd had clitoroplasties as babies (and who truly knew what had happened to them) said they were satisfied that that had been the right choice, and if most of those who'd been left with large clits regretted their parents' choice to forego infant cosmetic clitoral reduction surgeries, we would accept that infant cosmetic clitoral reductions worked to improve quality of life. In other words, we were clear that we were in this for people's well-being, not for some particular identity outcome.

This put us at odds with a lot of people in the movement. Many had come to see intersex as a core type of human identity, something that could only be solidified by surgery but never taken away. Bo had actively supported that identity formation; she had needed people to feel it to motivate them to fight. We didn't know of any successful rights movement that wasn't based on an essentialized shared identity (even if just constructed in politically expedient ways). Nevertheless, Bo and I decided we'd be perfectly happy if sex anomalies became so accepted that there simply was no intersex identity. We would be perfectly satisfied when the data showed that—with or without surgery—affected adults felt they had been treated justly. Our issue was not that funny-looking genitals held some special magical life-giving power that was being tragically taken away by surgeons. Our issue was not that hermaphroditic *identity* was being disappeared. Our issue was that women with big clits left intact seemed quite a bit better off than those who had been operated on. On the rare occasion when we met a woman with a big clit who had opted for surgery as an adult, she never regretted her parents' choice to leave it alone, and she always regretted her choice to have it shortened. We took that as further evidence that the problem was not identity as male, female, or intersex. It was the fact that the medical interventions didn't work: They didn't leave people better off.

Bo and I agreed that, if we put ourselves out of business—if, because of our work pushing for an evidence-based approach to intersex care, everyone born with a sex anomaly ended up feeling really great, so there was no need for an intersex sociopolitical identity, an intersex rights movement, or an ISNA—that would be just fine with us. We weren't in this for lifelong identities as intersex activists, as leaders of the "intersex community." The goal really was our goal. This again distinguished us, in ways I only later understood, from many activists, who bank on always being able to keep fighting over an identity issue. We *wanted* to retire. Our aim was to plant enough seeds of change in the medical system that change would continue without us.

Year by year, we saw more and more evidence of that possibility. In 2002, Jeffrey Eugenides' novel *Middlesex* came out. At first, it wasn't clear it was going to have much impact, but I knew when I looked at his book that I needed to pay attention to it. He credited my historical work in his acknowledgments, and his protagonist was a member of ISNA, the organization still legally registered to my home address. Very early in his book-tour cycle, I went over to one of Eugenides' readings at an Ann Arbor bookstore. I stayed after the public reading to talk to the author one-on-one. He worried aloud that no one wanted to buy this book—no one wanted to hear about intersex. I told him I thought that pretty clearly wasn't true, and that it would probably be OK. He alluded to the attributed shame he was getting for working on a book about this, and I sympathized, saying I was writing about conjoined twins and feeling like a freak. Look, I said to him, if we're feeling this much shame just by writing about this stuff, imagine what the people really living with it are experiencing.

The next time I ran into Eugenides was five years later, in 2007, at the Oprah studio, when Oprah featured *Middlesex* on her book club. By then, the book had been awarded a Pulitzer. (I was brought on to the show as the "medical expert" because they couldn't find any doctor who could explain intersex and sex development clearly to Oprah and her audience.) Not long after the 2002 Ann Arbor bookstore meeting, I could tell that Eugenides' novel was going to take off, because it seemed that on every flight I took, I was seated next to a person reading *Middlesex*. And it wasn't just young, gender-comfy people reading this book. It was old ladies and businessmen. (*Hermaphrodites,* I thought, *are a very marketable skill.*) And most important, doctors were reading this book, the doctors we were trying to change. I even heard from some who told me they were moved by this book in a way they had never been moved before—that they suddenly understood intersex to be one survivable part of a whole life. Here we had been feeding them real stories, but it was this novel that convinced them they needed to change their practices!

DURING THE BIG BLOOM in popularity of Eugenides' book, in 2004, Bo and I sneaked in together to the intersex session of the American Academy of Pediatrics Section on Urology, in San Francisco. We got a friendly doc to register us as "family" so that we could have name badges and slip in through security without being noticed on the regular registered participant list. We sat together and quietly listened to what the big guys were saying. They were tangled up in doubt. Progress!

Tina Schober—a surgeon out of western Pennsylvania who'd

become a pariah for associating with us—had actually been invited to speak. The UCSF surgeon Larry Baskin was admitting that they weren't really sure about where they should be cutting the clitoris, because the nerves were turning out to be located where they hadn't been expected to be, so that the outcomes were unclear, unpredictable. Indianapolis surgeon Richard Rink, who had always advocated "total urogenital mobilization"—ripping out everything that didn't seem right to the doctor and rebuilding a girl's genitals from scratch using Frankenstein stitches—was now expressing doubt about what the whole process was really based on. Of course, he then proceeded to say that, as a consequence, he was now just advocating "*partial* urogenital mobilization." But we were thrilled to hear Rink say that the most important consideration was "how to preserve function." He even told his colleagues, "I think there is a very important question: no one has proved it is a problem to have a large glans or a large clitoris, [so] should we really do anything about this?"

Afterward, we went up to say hello to the various panelists, and a few seemed shocked when they realized we had been there listening. I heard one person ask about what security thought they were doing at the door, but most of them were now cordial, though uncomfortable. Bo and I were not, after all, with the extremists on the picket line outside, anticircumcisionists who were covered in mock blood, calling doctors butchers. We were being taken very seriously.

However, because we were being taken seriously, we were taking crap from certain intersex activist quarters. I was an especially easy target of the identity-card-carrying activists. I was not intersex, I was not queer, I was not a clinician, so what was I doing there? In the story of the intersex rights movement, I was just plain funny looking. Some accused me of being a kind of mole—of "being in bed with the doctors." When people put this charge to my face, I asked whether they realized that being in bed with the doctors provided a lot more opportunities to tickle their nuts, so to speak, than simply yelling at them from outside the window. In fact, I admitted, I found the window-yellers useful, precisely because they made us look sane and reasonable. (I even donated cash to their groups to keep them yelling.)

And we *were* sane and reasonable.

But by then, totally exhausted. Eight years into our collaboration, Bo and I were both well on our way to being out of steam. When you think you're Good fighting Evil, you can continue fighting well past the point that would otherwise count as spent. But Bo and I had come to realize we were not Good fighting Evil. We were dealing with well-intentioned but myopic people who weren't seeing what we couldn't help but see when we took the long view in weighing the evidence.

Fully understanding how tired Bo was of it all, in 2004 I found myself having to push her to attend a hearing of the San Francisco Human Rights Commission that had been organized by David Strachan and Thea Hillman. The commission was investigating whether the treatment of intersex children constituted a human rights abuse. I thought this could be pretty scrumptious—I was fantasizing about writing our press release with the headline "Human Rights Abuses at U.S. Children's Hospitals." I talked Bo into going, and we went with Robin, Bo's wife. (Although at that time they could not be married legally because they were both identified by the state as women, Bo and Robin had decided a couple of years earlier to have a private wedding. They asked Aron and me to officiate.)

When it came time for public testimony at the Human Rights Commission, I got up and said a bunch of things, and so did other people, and everyone kept wondering, when would Bo get up and say something? When would the most prominent member of the movement speak? She finally realized she had to—there was no way out of what she saw as a waste of time. She'd told me the docs would never listen to the San Francisco Human Rights Commission, no matter what they said, so she had not prepared anything to say. By then I had developed the habit of leaving index cards in my bag so that I could quickly write down what I thought we should say. I grabbed some cards, started scribbling, and wrote something like this:

> What the Human Rights Commission has done here today is to recognize me as a human being. You've stated that just because I was born looking in a way that bothered other people doesn't mean that I should have been excluded from human rights protections that have been afforded other people. . . .

Bo got up and went to the microphone in the front of the little hearing room. Glancing down at the index card, she started to speak. And then she suddenly stopped speaking. I turned to Robin and, groaning, whispered, "Oh no, she can't read my handwriting."

Robin answered, taking my hand, "No, Alice, no. She's crying. She can read what you wrote, and she's crying."

Robin and I had never seen Bo cry in public before. I wasn't even sure I had seen her cry *ever.*

I realized at that moment that, after almost a decade together, we had finally gotten to the core of the matter. What I had learned from her was what I had written down on that card: that all she had ever wanted was simply to be treated as human. All of these people were simply asking to be given basic rights that were automatically accorded to all other humans: the right not to have your sex changed without your consent, the right to be told the truth about your medical history, and the right to be treated as an equal member of the human family without having to first pass through an operating theater.

We weren't asking for a new, third gender category for our society, nor for a belief in innate gender identity, nor anything else so culturally radical. We were just asking for children and adults who had been born with sex anomalies to be recognized as fully human, deserving of decent medical science, and deserving of basic human justice.

The good news was that a lot of people, including the doctors, were truly starting to get it.

CHAPTER 2

RABBIT HOLES

BO AND I managed to limp along together for about one more year after the Human Rights Commission hearing. In that last year of work together, we coordinated and published the first detailed consensus-based clinical guidelines for intersex pediatric care, along with a handbook for parents. By the time we finished, leaders from all of the major diagnosis-specific intersex support groups, clinicians from every relevant specialty, parents of affected children, and adult intersex activists all had signed on to the collaboration. Soothing these forty-some people into compromise over the phone for these texts damn near killed me. But when the two handbooks came out and were passed around in medical settings, even the old guard muttered appreciation. Although we lacked adequate data to know that our model was better for patients than the old way, we put forth an approach that seemed most likely to minimize harm, given what we knew historically and scientifically: For newborns with confusing sexual anatomy, assign a best-guess preliminary gender *label* of boy or girl, with the understanding that no surgery is required for a gender label. Provide medical and surgical care known to be needed to lower serious risk of illness or death, but hold off on all elective interventions, including elective genital surgeries, until the patient can decide. Provide ongoing psychosocial support by well-trained professionals for children and families. Above all, tell people the truth.

Right about the time we were getting ready to publish our handbooks, the big North American and European pediatric endocrinology groups decided to hold a conference on intersex care to arrive at their own "consensus." Bo was given an invitation to the meeting, as were several clinicians now firmly on our side. After talking with each ally who was invited, I put together a confidential list of talking points and gave it to each. As we hoped, a high-level international medical consensus emerged: The specialists agreed that they needed to work harder to collect and then follow long-term-outcomes data, to provide team care featuring dedicated psychosocial professionals, to find ways to tell patients and their parents the truth without making them feel overwhelmed and helpless, to stop counting patients who grow up gay or change their gender labels as medical failures, and to hold off on at least some genital "normalizing" surgeries until the patient could decide. Although these guidelines would not end surgical normalizations of genitals in early childhood right away, this consensus did mean that parents (and their sons and daughters) finally might get serious psychological support and be told what we know and don't know about intersex. Some of the doctors even started talking about shame, which had always been the real problem in intersex care. Moreover, they were *all* talking about needing to do better science to figure out what medical care really helps and what harms.

You'd think I'd have been dancing in the streets at this point, but like Bo, I was seriously worn down. For ten years, I had put up with the hardships of activism, and now the friendship with Bo that had long sustained me had started to evaporate. ISNA, once our joint baby, had morphed through its success into a sort of miserable small business, something neither of us felt especially excited about. A lot of pushing and pulling ensued. It turns out that having been through a war together doesn't necessarily mean you come back home able to make dinner together. I finally quit.

Relinquishing ISNA to Bo felt like losing a beloved stepchild in an unhappy divorce, and losing Bo as an intimate friend felt even worse.

It didn't help that just a few months before leaving ISNA—back when I was still kidding myself that I could keep working with Bo if I could just find a way to make my workload manageable—I had also quit my tenured professorship at Michigan State University. With a lovable four-year-old at my knee, I was tired of trying to do everything the university wanted of me. (A funny thing about writing manuals for parents of intersex children: You start thinking a lot about what's missing from your own parenting.) I just wanted to work from home, doing patient advocacy for victims of medical trauma, writing histories, and limiting my son's day care to six hours a day.

Then, like a tsunami after an earthquake, just a few months later, Aron was suddenly pulled from his medical-faculty position into an associate dean's chair at Michigan State, putting him essentially in charge of medical education at the university. While this meant plenty of family income to support me in my unconventional career move, it also meant that my rock of ten years had become the medical school's quarry just when I needed his grounding most.

Fortunately, not long after I'd turned in my resignation letter to Michigan State, a couple of colleagues had talked me into taking a part-time faculty position at their place, the Medical Humanities and Bioethics Program at Northwestern University's medical school in Chicago. The program's director promised that I could work almost entirely from home and basically do whatever work I wanted in exchange for putting the program's name on it. I could also have the unit's great faculty to lean on as colleagues. Still, with Aron suddenly absent, my job officially requiring almost nothing of me, and ISNA gone from my days forever, I found myself thoroughly unmoored—stumbling around as if I kept forgetting I'd had one leg amputated.

I found myself doing what any self-respecting straight woman does when she's disoriented by an identity-rocking emotional smash-up: I listlessly rearranged playlists and bookshelves while talking on the phone to my gay friend Paul. Paul Vasey is a Canadian scientist who spends part of the year studying the *fa'afafine*, biological males who live as women on the tropical island of Samoa, and part of the year studying girl-on-girl monkey action in the snowy mountains of Japan. (The *Weekly World News* once featured the macaque monkeys Paul researches under the headline LESBIAN MONKEY SHOCKER! Paul told me it was actually a pretty good article.) Not long after the time I was calling him three times a week for company, Paul and his colleagues conducted a formal study of "fag hags"—straight women like me who have many gay male friends. They showed scientifically what Paul demonstrated in my life that year: that gay men make their close women friends feel better about themselves. Being a hard-driving scholar like me, Paul knew—and told me bluntly—that I just needed a big new project, one that would feed my hungers for intensive historical research and social justice. Soon enough, he'd lead me into one: the Bailey transsexualism controversy.

WHEN PEOPLE ASK ME how transgender is different from intersex, I usually start by saying that intersex and transgender people have historically suffered from opposite problems for the same reason. Whereas intersex people have historically been subjected to sex "normalizing" hormones and surgeries they have *not* wanted, transgender people have had a hard time getting the sex-changing hormones and surgeries they *have* wanted. Both problems arise from a single cause: a heterosexist medical establishment determined to retain control over who gets to be what sex.

Aside from that huge shared problem, intersex and transgender actually are quite different. By definition, intersex involves having some anatomical feature that makes one's body atypical for males or females; it's primarily about anatomy—your body. By definition, being transgender means rejecting the gender assignment that was given to you at birth; it's primarily about self-identity—your feelings. Although a small minority of intersex people do reject their birth gender assignment, most don't, and most transgender people weren't born intersex. In the great majority of cases, medical scans won't detect any intersex feature in a transgender person's body. Nevertheless, many people *believe* that transgender must be a special form of intersex involving the brain. Here's that popular, comforting narrative: Everyone is born either male or female in the brain. But a person might accidentally be born with the "wrong" sexual anatomy—be born with an essentially female brain in a male body, or vice versa. If this happens, the person will know from early childhood that a terrible mistake has been made. If fortunate, such a person will eventually be able to come out of the closet and use surgery, hormones, and the legal system to end up with the body and social identity she or he should have always had.

Although there is very little science to support it, this has become the most popular explanation of transgender, probably in part because it is the easiest one for uptight heterosexuals to accept. Some people *appear* to switch sides, but everyone can rest assured that they didn't *really* switch; they just finally got sorted out correctly by having their internal gender realities externalized by transsexual hormone treatments and surgeries. In practice, this story of transgender can function as a kind of get-out-of-male-free card for men who seek to become women anatomically. When that card is played, the comforting narrative of "true selves" is preserved. Everybody *really* has just one true gender from birth to death, so gender seems ultimately very stable. Now, no one really gets out of being male for free—the physical, financial, and personal costs of transition are pretty high—but this narrative does give a person a way out to which other people can't easily object, at least in America, where the quest for the true self counts as admirable, even sacred.

If Northwestern University psychology professor J. Michael Bailey had accepted this story of male-to-female transsexualism, he and I might never have met, because he never would have gotten himself into such a pickle with transgender activists. But as I was to learn, Mike Bailey has never cared for simple, politically correct stories. In fact, he liked using his research and his college classes to kick politically correct assumptions around until they were as dented as soda cans on the sidewalk. In his view, the simple "female brain in a male body" was unscientific and had to go.

In 1997—right about the time I had started helping Bo with ISNA—Bailey decided to write a book for the general public about "feminine males." His decision came after he attended a Barnes & Nobel book-signing by a Chicago-based therapist named Randi Ettner, who was promoting her new book, *Confessions of a Gender Defender.* In it, Ettner pushed the politically correct "brain of one sex trapped in the body of the other" story of transgender. This rankled Bailey. Make no mistake: It wasn't that he wanted to stop transgender people from getting access to the hormones and surgeries they wanted. Far from it. As a libertarian, he always wanted to see these folks get whatever medical technologies they needed to feel whole, just as Ettner did. But he also wanted to replace what he saw as a false picture of male-to-female transgender with what he saw as the true one. He wanted better science *and* progress for transgender rights, and he hoped to help push both by writing his own popularization.

It took Bailey another six years, until 2003, to complete and publish *The Man Who Would Be Queen: The Science of Gender-Bending and Transsexualism.* The first hint that this work would reject simplistic gender-identity stories—from transgender people or anyone else—came from the book's provocative and insensitive cover. It featured a photo of two hairy masculine legs standing in a pair of pretty pumps, shown from the knees down—an image seen by most people (including me) as more befitting a Monty Python cross-dressing sketch than a book about science from a trans-friendly writer. (Bailey chose this cover against the advice of colleagues, who preferred a vastly less offensive alternative showing three faces, one feminine, one masculine, and one androgynous.) Meanwhile, in the text, like a lot of feminists with whom he otherwise tangled, Bailey rejected the idea of anybody being simply male or simply female in the brain. He suggested to his readers that "gender identity is probably not a binary, black-and-white characteristic. Scientists," he complained, "continue to measure gender identity as 'male' or 'female,' despite the fact that there are undoubtedly gradations in inner experience between the girl who loves pink frilly dresses and cannot imagine becoming a boy and the extremely masculine boy who shudders to think of becoming a girl."

Rejecting the idea that everybody is truly and easily assignable to one of two gender identities, Bailey unapologetically and aggressively introduced his readers to a generally unfamiliar understanding of male-to-female transgender. This understanding depended not on an idea of a "true female" trapped within, but on sexual orientation. Male-to-female transgender, in Bailey's view, was more about *eroticism* than gender identity per se. Here Bailey was drawing on the work of Ray Blanchard, a sex researcher working at Canada's Centre for Addiction and Mental Health in Toronto. When Blanchard considered the historical and clinical literature and his own experience working as a psychologist with hundreds of adult men seeking sex reassignment, he realized that there are *two* basic types of male-to-female transsexuals, very different from each other in terms of their life histories and demographics. Blanchard also realized that these two types could be recognized primarily by their sexual orientations. Blanchard concluded that male-to-female transsexualism isn't simply about gender identity (whether you really feel yourself to be male or female) but is fundamentally about sexual orientation (whom or what you really desire).

The first of the two types of male-to-female transsexuals identified by Blanchard begin life as very femme little boys. They are "sissy boys" who like activities generally considered girly. Little boys of this type like to dress up in girls' clothes, play house, and play games involving fashion, and many are downright averse to "boyish" rough-and-tumble sports or games of war. They seem more classically feminine than masculine in social interactions. They are highly attuned to social relations and often like helping their mothers with housework. From the moment of the development of their sexual interests, these folks are unequivocally sexually attracted to other males. Before they become women, in their behavior and their exclusive sexual interest in men, they appear to be superfemme gay men. Unfortunately, this means that their sexual opportunities are often limited while they are presenting themselves as men. Straight men aren't interested in having sex with them because they're male, and gay men often aren't sexually attracted to them because most gay men are sexually attracted to masculinity, not femininity, and these guys are really femme.

Now, because they are so femme, when they dress as *women*, even before hormonal or surgical transition, these folks pass pretty easily (as women), which means straight men become interested in them. Needless to say, they pass even better after hormonal and surgical transition. As post-transition women, they can and happily do take straight men as their sex partners. As a consequence, for these male-to-female transgender women, sex reassignment makes possible a more satisfying sex life and a more comfortable gender presentation, as they no longer have to fight to dampen their natural femininity. Transition also means a less painful and safer life; as women, they are not as often subject to homophobic abuse and assault, always a danger for femme men.

In articulating this demographic profile among the clinical population of male-to-female transsexuals presenting for sex reassignment, Blanchard called these people homosexual transsexuals, because they are natal males sexually attracted to other males. Although scientifically precise, this is a term I generally find confusing because, when these people change their sex from male to female, they then become heterosexual. Therefore I generally use the less confusing and less clinical term that many of these folks use for themselves—*transkids*, a term that recognizes that in this population obvious gender boundary crossing starts early in life.

Bailey's book contained portraits of several transkids, but the most vivid was that of an attractive Latina trans woman identified as Juanita. Juanita came across in Bailey's prose as a seriously attractive and highly sexed woman, one who made a good living for a time as a sex worker. After her transition, she eventually landed a nice husband in the suburbs. However, Bailey reported, she finally gave all that up because she missed the excitement of the city—including the sexual variety. Juanita perfectly embodied Bailey's understanding of transkids: She was a sexy, very feminine individual with a typically male (strong) interest in sex and a typically male (high) interest in casual sex—the kind of person, Bailey said, who "might be especially well suited to prostitution" (groan).

OK, now here drops the other shoe, the part of Blanchard's theory that really pissed off the people who went after him: Blanchard noted that nonhomosexual transsexuals—putatively straight, bisexual, and asexual men asking for sex reassignment—looked very different from the transkids in their life histories, their habits, and their sexual interests. Members of this second group were *not* markedly femme in childhood. In fact, as kids they seemed to everybody like typical boys—often having been into sports, military play, and vehicles. Many of them went into fairly male-typical occupations—they were engineers, mathematicians, and scientists in heavily quantitative fields. They were exclusively or primarily sexually attracted to females, and before transition they were typically married to women and had fathered children. Unlike the transkids, these men looked to the outside world like typical straight men right up until transition. People who knew them were usually completely shocked when they announced they were going to change sex. But they didn't *feel* like typical straight men do. Unlike most straight men, they felt there was a powerful, almost overwhelming feminine component of their selves. Part of that sense involved finding themselves sexually aroused by the idea of being or becoming women.

Wait—what? What does that last bit look like? Well, in terms of fantasies, it's a variation on what each of us fantasizes about sexually, when masturbating, for example. One person might imagine oral sex with a particular movie star. Another might imagine being tied up and teased by a stranger. Another might imagine "plain vanilla" intercourse with a lover. Note that all these sexual fantasies involve gender; that is to say, when we fantasize sexually, we typically do so in ways that specify the genders of the parties involved. Gender can be seriously pleasurable for most of us. For the man who is sexually aroused by the idea of becoming a woman, the gendered component of the fantasy is brought to the fore; the simple idea of being or becoming a woman causes sexual excitement. Blanchard coined a new term for this type of male-to-female transsexualism: *autogynephilia*, meaning self-directed (auto) love of females (gynephilia). Pretransition, these men experience "love of oneself as a woman," a phrase downright beautiful in French: *amour de soi en femme.*

Blanchard's taxonomy of male-to-female transsexuals recognized the importance of sexual orientation in the gendered self-identities of both those who begin as homosexual males and those who experience *amour de soi en femme.* However, he didn't see sexual orientation as the *only* thing a male factors in when deciding whether to transition. He recognized that in one environment—say, an urban gay neighborhood like Chicago's Boystown—an ultrafemme gay man might find reasonable physical safety, employment, and sexual satisfaction simply by living as an ultrafemme gay man. But in a very different environment—say, a homophobic ethnic enclave in Chicago—he might find life survivable only via complete transition to womanhood. Whether a transkid grows up to become a gay man or a transgender woman would depend on the individual's interaction with the surrounding cultural environment. Similarly, an autogynephilic man might not elect transition if his cultural milieu would make his post-transition life much harder.

To firmly make this point about biology interacting with environment, Bailey's book begins and ends with the story of Danny. Bailey presents Danny at the start of the book as a femme little boy who seemed to want to be a princess and whose mother consequently came to Bailey for advice. At the end of the book, the reader learns that Danny has been growing up into a young gay man in the care of accepting and loving parents. Surprisingly—given that the author is a self-identified genetic essentialist—Bailey used Danny's story as the frame for his book to teach the importance of *culture* to identity. Bailey wanted the reader to understand this: Although people are born inclined to particular sexual orientations and gendered behaviors, transgender isn't something you're born either to be or not to be. Contrary to the popular mainstream understanding of transgender, whether you end up living as an out gay man, a closeted gay man, or a straight transgender woman depends not only on biology, but also on *cultural tolerance* of various identities. Biology plus cultural environment equals the experience of identity.

Just to be clear: This interplay of biology and culture affects everyone's experience of identity, not just transgender people's. To take an example familiar to most women, whether it matters that my female physiology turns my occasional frustration into tears rather than punches, whether I feel denigrated or accepted as a person hormonally prone to involuntarily crying—that depends on the culture around me. So, while in a very homophobic environment Danny might have grown up to be a transgender straight woman, because his parents and community generally accepted that he would always be a male attracted to men, he was growing up instead to be a gay man.

Paul Vasey has been documenting that in Samoa little boys who are naturally very femme are welcomed into a special gender category called *fa'afafine*, a term that signifies living "in the manner of a woman," and are raised like the girls. The *fa'afafine* grow up in female roles, tending to the family and taking men as their lovers, although they almost never alter their anatomy, because their traditional culture doesn't require them to do so in order to live as women socially and sexually. They are transgender women without any hormone treatments or surgery. The *fa'afafine* are understood and accepted this way by their families and their lovers, and have been for generations. In now-Christianized Samoa, they wouldn't be so well understood or accepted if they self-identified as gay men.

Samoa is not unique; indeed, Bailey's book pointed out that "homosexual transgenderism" is the most common form of transgenderism around the world, found, for example, in Thailand, Mexico, Iran, and Albania. In many cultures, homosexual transgenderism has functioned for countless generations as a way to "straighten out" homosexual desires—a response that can be tolerant and progressive (as in Samoa) or repressive (as in Iran), depending on how it is enacted and experienced.

Thus, whether transkids change sex medically and surgically depends on the way their biology, their psychology, and their culture interact. As Blanchard and Bailey noted, the same is true for the men who become women by way of *amour de soi en femme* (autogynephilia). Whether a man who dreams erotically of becoming a woman opts to change sex hormonally and surgically depends upon the interaction of the individual's body and psychology with the local cultural environment. Some of these folks simply fantasize about crossing into womanhood while remaining apparently typical straight men socially their whole lives. They may try to suppress the thoughts, or they may enjoy transsexual erotica, but they limit themselves to thoughts and dreams. Others will occasionally cross-dress for erotic purposes and to enjoy temporarily experiencing a deep feeling of femininity. The psychiatrist Richard/Alice Novic writes about this kind of bi-gendered life in the autobiographical *Alice in Genderland.* Novic spends part of the time living as Richard with his wife and children and part of the time as Alice, enjoying the company of a boyfriend who accepts and desires her as she is. Unlike Novic, a few men who experience *amour de soi en femme* do decide to seek medical interventions. Some opt to get breast implants or to take female hormones—to enhance their sense of being a woman and to be socially "read" as women—while keeping their male genitals. And a few autogynephilic individuals find that they can feel fulfilled only by complete hormonal, surgical, and social transition to women. For those who seek full transition, that's what it takes to feel that they are living an authentic life, true to themselves.

In Bailey's book, the portrait of *amour de soi en femme* male-to-female transgenderism came in the form of a trans woman identified as Cher, née Chuck. The reader gets the sense that Bailey likes Cher, but his portrayal of her is startling. As a boy and young man, Chuck seemed like a pretty typical guy. But unlike most guys, Chuck made elaborate pornographic films of himself dressed as a woman, complete with female masks, homemade fake breasts, and a glue-on vulva. (He pushed his penis up into his body and used an adhesive to hold it there, a feat made easier by having been born with only one testicle.) Chuck made robots to simulate heterosexual sex, so that he could experience sex as he thought a woman might. Eventually he realized he needed to be seen by all as the woman felt inside. And so, with the help of gender-affirming hormones and surgery, Cher emerged.

In Bailey's account, Cher blossomed after transition, finally able to live out the gender identity she had long felt and desired. As is the case for transkids like Juanita, transition can make the lives of people like Cher far more fulfilling. It lets them be who they feel they really ought to be, who they really *are.* Life is surely a lot easier when people treat you the way you feel you should be treated in terms of your gender identity and sexual orientation. Perhaps not surprisingly, women from both groups—Juanita's and Cher's—report on average doing better psychologically post-transition. Blanchard is one of the researchers who has documented substantial improvement for well-screened trans women, and he has used that outcomes research to openly support, in sworn testimony, the public funding of sex reassignment in Canada. Blanchard and Bailey see trans women who begin with homosexual desires and those who begin with *amour de soi en femme* as radically different from each other in some ways (childhood gendered behaviors and sexual orientation) but equally deserving of quality care, human rights, and public support. But the subtlety of their position—supporting political rights for transgender people while promulgating a politically incorrect reading of male-to-female transgender—was lost on certain transgender activists, who attacked Bailey for his support of Blanchard's theory.

Still, an outsider might wonder how Bailey's acceptance of Blanchard's conclusions could be read as so very offensive. Why were so many trans people willing to enlist in the army assembled to fight Bailey within days of his book's publication? In the first pages of his book, Bailey summed up Blanchard's taxonomy of male-to-female transsexuals in rather tender prose: "Those who love men become women to attract them. Those who love women become the women they love." It sounds so sweet, really, all that love. But most of Bailey's book represented male-to-female transsexuality as a matter of *lust*—gay lust in the case of the transkids like Juanita and self-lust in the case of autogynephiles like Cher. When Bailey talked about people like Cher, he explicitly labeled their sexual desires *paraphilic*—a word that to many means "sick" even when it refers to consensual, nonexploitive sexual interests.

Therein lay a real problem, one that explains why the transgender activists who went after Bailey were able to garner fairly widespread help from other transgender people, at least at first. Before Bailey, many trans advocates had spent a long time working to *de*sexualize and *de*pathologize their public representations in an effort to reduce stigma, improve access to care, and establish basic human rights for trans people. The move to talking about trans*gender* instead of trans*sex* was motivated in part by a desire to shift public attention away from an issue of sexual orientation (sexuality always being contentious) to an issue of gender. This is similar to how gay rights advocates have desexualized homosexuality in the quest for marriage rights, portraying themselves in living rooms and kitchens instead of bedrooms, in order to calm fearful heterosexuals.

The de-emphasis of sexuality among trans advocates also occurred because some of the sexologists and clinicians who had acted as gatekeepers for sex reassignment had for many years maintained the heterosexist idea that the transkids like Juanita were the only good candidates for sex reassignment. The transkids were naturally pretty femme and could be counted on to remain stably attracted to men, so they passed easily as straight women with no massive disruption of the heterosexual order. Meanwhile, many applicants for reassignment who showed any hint of *amour de soi en femme*—the would-be Chers—found they had to lie about their orientations and histories to get what they needed. Otherwise some clinicians were reluctant to let them get the hormones and surgeries they sought. In 1969, one clinician had indicated that a *single instance* of arousal by cross-dressing should eliminate a man from candidacy for sex-reassignment surgery.

Indeed, a few retrograde clinicians, like Paul McHugh, a psychiatrist at the Johns Hopkins University School of Medicine, still actively use the idea that male-to-female transgender is really about perverted sexuality and mental illness to argue against access to sex-transitional hormones and surgeries. McHugh and his ilk ignore the fact that well-screened trans women are better off after transition, as if their well-being and happiness are utterly irrelevant. McHugh (whom Bailey actually criticizes in his book) has likened sex reassignment to doing liposuction on anorexics, apparently not noticing that, um, anorexia *kills people* while sex reassignment for adults *saves lives*. All this has contributed to the mainstream trans community's feeling that it makes more sense to emphasize issues of gender identity rather than issues of sexuality. After all, under any reasonable understanding of human rights, one's sex life ought to be one's own business so long as one isn't hurting anyone else.

It's also worth mentioning, given how often cultural sex politics play out in universities, that academic feminists have always seemed a lot more supportive of trans*gender* than trans*sexuality*—that is, when they've been supportive at all. The history of feminism and trans issues has been fraught with tension, especially since feminist Janice Raymond's 1979 book, *The Transsexual Empire*, which accused trans women of actively undermining the work of "real" feminists by supposedly giving in to the heterosexist patriarchy by simply switching over from stereotypical male to stereotypical female. Raymond even claimed, "All transsexuals rape women's bodies by reducing the real female form to an artifact, appropriating this body for themselves." Some feminist groups, like the Michigan Womyn's Music Festival, still shun transgender women, admitting only "womyn born womyn." (So much for Simone de Beauvoir's observation that women are not born, but made.)

Add to *that* the fact that, in many places, discrimination against trans people has been perfectly legal in housing, employment, even schooling. Then add the history of police refusing to investigate (or even participating in) gay-bashings and murders of trans people, not to mention emergency workers refusing to treat trans people with life-threatening injuries, and you get a group understandably vigilant about possible violations of their rights.

In short, there is always a lot at stake politically and socially when you're talking about transgender. And yet, while some of Bailey's best friends really were gay men and trans women, in his clueless privileged way, he didn't worry about his work's political implications for sexual minorities. He worried only about what's *right* scientifically, and he decided that Blanchard's taxonomy was right about the salience of sexual orientation to male-to-female transsexuality. Bailey gave quite sympathetic portrayals of all the trans women in his book, including Juanita and Cher, and he firmly concluded that the ultimate happiness of individual transgender people is what matters most, even if transitions leave families or communities unhappy. But Bailey made the mistake of thinking that openly accepting and promoting the truth about people's identities would be understood as the same as accepting them and helping them, as he felt he was. Where identities as stigmatized as these are concerned, it just isn't that simple. The shame and derision accorded trans women like Juanita and Cher doesn't disappear just because a few scientists may be *personally* fine with the idea that men might become women primarily because of reasons of sexuality, not "trapped" gender identity. As I came to learn, Bailey thought sexuality was a plenty good reason for lots of actions. But the trans women who attacked Bailey for his book understood that the world would probably not agree.

And they weren't interested in finding out. They wanted the whole business of Blanchard's taxonomic division shot down. Transsexuality should appear only as the public could stomach it, as one simple story of gender, a tale of "true" females tragically born into male bodies, rescued and made whole by medical and surgical sex reassignment. And there should be absolutely no mention of autogynephilia or any other sexual desires that might make trans women look to the sexually sheltered like the perverts they were historically assumed to be.

To understand the vehemence of the backlash against Bailey's book, you also have to understand one more thing. There's a critical difference between autogynephilia and most other sexual orientations: Most other orientations aren't erotically disrupted simply by being labeled. When you call a typical gay man homosexual, you're not disturbing his sexual hopes and desires. By contrast, autogynephilia is perhaps best understood as a love that would really rather we didn't speak its name. The ultimate eroticism of autogynephilia lies in the idea of *really* becoming or being a woman, not in being a natal male who desires to be a woman. At least in fantasy, the typical autogynephile erotically desires a complete identity transformation—*to be a woman*, not to be a transsexual. So when the autogynephilic psychiatrist Richard/Alice Novic talks about her boyfriend stimulating her genitals, she refers to her "clitoris," although anatomically what she has is a penis. The erotic fantasy is to really be a woman. Indeed, according to a vision of transsexualism common among those transitioning from lives as privileged straight men to trans women, sex reassignment procedures are restorative rather than transformative, because the medical interventions "fix" what some call the "birth defect" in their natal bodies. Some even reject the label of transgender or transsexual for this reason; to themselves they are simply women, outside now as they always were inside.

For Bailey or anyone else to call someone with *amour de soi en femme* an autogynephile or even a transgender woman—rather than simply a woman—is at some level to interfere with her core sexual desire. Such naming also risks questioning her core *self-identity* in a way that calling the average gay man homosexual simply can't. One really must understand this if one is going to understand why some trans women came after Bailey so hard for naming and describing autogynephilia. When they felt that Bailey was fundamentally threatening their selves and their social identities as women—well, it's *because he was*. That's what talking openly about autogynephilia necessarily does.

In a nutshell—and this is really indisputable—it was Bailey's dangerous dissemination of this part of Blanchard's work that led a prominent transgender woman named Lynn Conway to begin what became a war against Bailey from her base at the University of Michigan, where she was on the computer engineering faculty. As Conway must have understood, Blanchard's scientific work, always written in rather dry prose and published in hard-to-access specialist journals, could never pose the threat of Bailey's *The Man Who Would Be Queen*, with its intriguing scope, engaging prose, sex-positive tone, and compelling personalized portrayals. Bailey's book constituted a serious innovation. It could well bring Blanchard's taxonomy—including news of autogynephilia—to the masses and change the public perception of women like Conway.

And so, within days of publication of *The Man Who Would Be Queen*, Lynn Conway sent an urgent e-mail to a trans woman ally named Andrea James:

> I just got an alert about J. Michael Bailey's new book. It's just been published and of all places it's co-published by the National Academies Press, which gives it the apparent stamp of authority as "science." . . . As you may know, Bailey is the psychologist who promotes the "two-type" theory of transsexualism. . . . Anyways—not that there is much we can do about this—but we should probably read his book sometime and be prepared to shoot down as best we can his weird characterizations of us all.

You're probably wondering how I got that e-mail. The answer is that Conway developed what became an enormous Web site hosted by the University of Michigan for the purpose of taking down Bailey and his ideas. There she proudly and steadily recorded her efforts against Bailey, Blanchard, et al. In fact, it was her own university Web site that largely enabled me to figure out what she had really done and how Bailey had essentially been set up in an effort to shut him up about autogynephilia.

WHEN BAILEY'S BOOK emerged in 2003, I didn't pay much attention to the mushroom cloud expanding over Evanston, Illinois, where Bailey was tenured in Northwestern's Psychology Department. At that time, I was still working on intersex at Michigan State. But people I knew were increasingly trying to get my attention focused on it. Paul Vasey kept telling me he couldn't believe what transgender critics were doing to Bailey and even to his children and girlfriend, and Lynn Conway herself was calling me to help go after Bailey. In fact, as I found out via Paul, Conway had simply added me to her Web site's list of outraged allies, apparently assuming I would agree with her.

But in 2003 I waved both Vasey and Conway off. To Paul, I said I didn't have it in me to feel sorry for a member of the sexology establishment, given what the bastards had done to intersex children. To Conway I was more polite. She was a major donor to ISNA, and we didn't have a lot of those. Still, I told her to take my name off her list of Bailey opponents—I hadn't even read the book—and I advised her to just ignore Bailey. Paying attention to him, I told her, will just help sell his book.

Conway and company didn't give up, however. The mounting pile of national press reports on the scandal made that clear. Whereas at first the complaint was that Bailey's book portrayed a wrong and offensive vision of men who seek sex changes, soon the complaints became less about his supposedly offensive theory and more about his allegedly unethical actions. The emerging charges looked bad: that Bailey had failed to get ethics board approval for studies of transgender research subjects as required by federal regulation; that he had violated confidentiality; that he had been practicing psychology without a license; and that he had slept with a trans woman while she was his research subject. The wildfire nature of the conflagration made me no more inclined to get near it.

In early 2006—three years after Mike Bailey published his book, just after I quit ISNA, and a few months into my Northwestern position—I made plans to meet Paul in Chicago. We had decided to edit together a special journal issue on the evolution of sex, so we were meeting to hash out the details. But Paul also wanted to use the trip to introduce me to Bailey. Paul had been telling me about him off and on for years, mostly to tell me about the hell he'd been going through at the hands of the transgender activists. Paul said that the whole experience had terrorized Bailey, that he was a different man than before the controversy.

By that time, at Paul's request, I had read *The Man Who Would Be Queen*. Paul had always insisted that it was a very good book about the range of feminine males. After a careful read, I had responded to Paul that it *was* certainly very original and engaging—and much more explicitly supportive of gay and transgender rights than I had expected—but that it had some truly obnoxious parts in it. Granted, they amounted to just a few lines, but they grated. For example, there was the bit where Bailey claimed he could "know" much about the childhood and sexual orientation of a man he had just met, merely because the man was quite femme; the line where Bailey called one entire group of trans women (those with *amour de soi en femme*) not particularly good-looking; and that passage where he suggested that the other group (the transkids) might be especially well suited for sex work because after transition they supposedly possess the perfect combo of traits and interests. There was also one place where Bailey wrote this: "When [the transkid Kim] came into my laboratory, my initial impression was reconfirmed. She was stunning. (Afterward my avowedly heterosexual male research assistant told me he would gladly have had sex with her, even knowing that Kim still possessed a penis.)" I got that in this passage Bailey was trying to convince people to get over their knee-jerk transphobia—but *honestly,* calculating a trans woman's attractiveness using this particular metric seemed a bit much for a book by a scientist. Then there was the dreadful cover photo. I didn't know if the nasty things people had been saying online and to reporters about Bailey were true. But I knew it *was* true that the book read as if written from a place of heterosexual white male privilege even though it was surprisingly saturated with genuine affection, sympathy, and unabashed support for femme boys, gay men, drag queens, and transgender women.

In advance of meeting in Chicago, with my tense approval, Paul set up dinner for the three of us in Boystown, just a few blocks from Bailey's home. Over appetizers and then dinner, in spite of Paul's attempts to get us to warm up, Bailey and I treated each other with cool suspicion. Yet it quickly became obvious to me that Bailey had great affection for Paul, and vice versa, and I just couldn't figure out what to make of this. Paul was generally a good judge of people. Was Bailey like one of those characters in a novel who is nothing like his public reputation?

It was also obvious that Bailey was completely comfortable with the gay men all around him—including, I soon realized, one of my former students from Michigan State, who was at the next table with his boyfriend and who threw his arms around me and kissed me when he realized it was me yammering away right next to him. Unlike many straight men I knew, Bailey did not seem awkward around so many gay men, nor was he trying hard to prove that he was comfortable. He just *was* comfortable. I recalled that in his book he'd said that the problem isn't that gay men are, on average, more feminine in their interests and behaviors than straight men; the problem is that we think that's a *bad thing* and try to deny it rather than accepting it—accepting *them.* Femiphobia, he called it. (Others have called it sissyphobia, a term that I think better captures the problem.) Sitting there, I realized that maybe he was right about lack of acceptance being the problem, because he seemed perfectly at ease with the great variety of masculinity—heavy and light—around us that evening. He seemed like the future of straight people's full acceptance of gay people.

At some point during that dinner, utterly disoriented, I asked Bailey point-blank if he had slept with a trans woman research subject—the most scandalous of all the charges made against him. He looked very tense and launched into what sounded like a canned legalistic response, saying that, even if he had (and he wasn't saying whether he had), there is nothing automatically wrong with having sex with a research subject.

Say *what*? I was intrigued. After dinner, we three walked over to Sidetrack, a gay bar a couple of doors down, and when Paul wandered off into the crowd for a few minutes, I told Bailey I was sorry they had gone after his kids. He just said, "Thank you," and had another drink.

I couldn't figure this guy out. How could someone so soft-spoken get into so much trouble? Why would someone so very polite and politically progressive write those few really obnoxious lines in his book? Could it be, perhaps, that he was not homophobic or transphobic (his book certainly wasn't, nor did he seem to be) and not tone-*deaf* but merely tone-*dumb*? Maybe he was someone who could *hear* the political music around him very well but lacked the ability to sing along in tune. His book *was* rather like a generally elegant solo performance punctuated by a number of teeth-grinding sour notes.

In the next couple of days, I poked around a bit. I looked up Bailey's work and saw that most of it consisted of serious peer-reviewed scientific articles, quite different from his chatty and footnote-free book. I came across his old twin studies—controversial work that had showed that identical twins are more likely to have the same sexual orientation than are siblings who are not genetically identical, strongly suggesting that sexual orientation may sometimes be inborn. Suddenly I placed his name: I had actually taught my undergraduates criticisms of

Bailey's work on twins many years before. Back in those days, saying gay people might have been born that way, as Bailey was doing, was politically *un*popular among many gay-rights activists and among humanists in the academy, who were fighting any claims of unalterable or predestined "human nature." Back then, too, Bailey had sounded proudly tone-dumb.

I dug a little more. Knowing that one of Bailey's book's critics had claimed Bailey had "abandoned" his wife and children, I took a close look at the personal information portion of his Web site. The Bailey clan appeared to be one of those post-divorce families that is still fundamentally a family. If Bailey was faking that, it was a convincing fake, but his critic's claim that he had abandoned his wife and children had been very effective in skewing this wife and mother's impression of him.

As I kept digging, I noticed something even more interesting: Many of the trans people whose scholarship and political work I had most admired in the last ten years had remained strangely silent in the Bailey controversy. They had apparently steered clear. As had I.

THAT WAS FEBRUARY 2006. In May I got an e-mail from Mike Bailey bemoaning the fact that Northwestern's Rainbow Alliance, our university's LGBT group, had invited Andrea James to speak. I confessed to Mike that I had never really sorted out the characters in his controversy and asked him to remind me who she was. He sent me a PDF documenting how, in 2003, Andrea James had downloaded pictures of Bailey's two children, Kate and Drew, from Bailey's Web site and put them up on her own site, www.tsroadmap.com. When the photos were taken, Kate was in elementary school; Drew, in junior high. James had blacked out the children's eyes, making them look like pathology specimens, and asked in a caption below, whether Kate was "a cock-starved exhibitionist, or a paraphiliac who just gets off on the idea of it." The text went on to say that "there are two types of children in the Bailey household," namely those "who have been sodomized by their father [and those] who have not."

I was pretty stunned. Others had told me about this tasteless stunt, but I had never seen it for myself. It was obvious James was trying to parody Bailey's book, but to what end? What kind of person undermines a rights movement by using this kind of creepy tactic?

So I promptly wrote to the Northwestern Rainbow Alliance, first apologizing that I hadn't previously introduced myself. I explained that being a long-distance part-timer based on the Chicago med school campus meant I had almost no interaction with the Evanston campus, where the Rainbow Alliance (like Bailey) was housed. I offered to speak sometime about intersex, and then got to why I was writing, namely to register my protest at James being invited to the campus. I said I didn't think she was good for a scholarly institution, nor did I think she was good for trans rights. They didn't answer. Frustrated, on the day before Mother's Day in 2006, I blogged about this on my personal Web site. Knowing a bit about James's tactics, I called the essay, "The Blog I Write in Fear."

Behold the floodgates opening. Now a few people from the Northwestern Rainbow Alliance did write back to me, to take issue with my criticism of their decision, and several trans women did the same. Meanwhile, fan mail arrived from a number of sex researchers and from Bailey's daughter, Kate, now a college student. Some trans women wrote to tell me that no matter how Bailey was wronged, he deserved whatever he got. A couple more trans women wrote to me that Bailey was *right* about them all, and James knew it—that *that* was the problem.

But the most interesting mail, from my perspective, came from trans women who wrote to tell me that, though they weren't thrilled with Bailey's oversimplifications of their lives, they also had been harassed and intimidated by Andrea James for daring to speak anything other than the politically popular "I was always just a woman trapped in a man's body" story. They thanked me for standing up to a woman they saw as a self-serving bully.

In what in retrospect seems like a stupid move, I also made a point of writing to Andrea James to tell her about my blog and to suggest that she tone down her rhetoric lest she undermine the trans-rights movement. Oh, she didn't like that. She didn't like that one bit. She wrote back a series of nasty e-mails, including one referring to my son as my "precious womb turd." (Paul soon took to asking after "the precious womb turd" when he called.) She also showed up at my office when she was in Chicago, leaving her card in my mailbox. Then she e-mailed me, subject line "Mommy Knows Best," saying, "Sorry I missed you the other day. Your colleagues seem quite affable, and not as fearful as you. . . . Bad move, Mommy." She closed, "We'll chat in person soon." My dean suggested I talk to university counsel, who asked that I check in with the university police.

Now that I'd learned a little more about one of Bailey's chief critics, I knew I had to investigate this controversy. Now I really wanted to know what was going on here.

IT SOUNDS FUNNY TO SAY, because I had read Bailey's book years before I met him, but it was only when I read it again alongside Blanchard's papers, in order to start understanding the history of the controversy, that I truly became fluent in the division of male-to-female transsexuals into those who begin with homosexual desire and those who begin with *amour de soi en femme*. And as I did, the lives of trans women I knew personally suddenly started to make more sense. In fact, I now found one prominently featured section of Lynn Conway's Web site—"Photos of Lynn"—sort of ironically funny. Here was this woman dedicating most of her life, it seemed, to attacking the concept of erotic arousal from the idea of being a woman as the basis for one form of male-to-female transsexualism, while simultaneously putting up—on *her university Web site*—multiple pictures of herself in a skimpy bikini, shot from various angles. In addition, there were pictures of Professor Conway in miniskirts, in a little black dress, and in her white bridal gown. As if that weren't enough, Conway gave her measurements (41-32-41) and did not neglect to mention that her hair is light brown/auburn and her eyes are blue. Just your average computer engineering faculty Web site, nothing sexual, right?

But what astounded me even more than Conway's Web pages was evidence that—before Conway had called them to arms—Conway's two chief compatriots in the assault on Bailey's reputation had pretty much acknowledged that they had been sexually aroused by the idea of being or becoming women. One of those two was Deirdre McCloskey, a distinguished professor of economics and rhetoric at the University of Illinois, a woman who fell in as the third musketeer to Conway and James. As I started to figure out via my roughed out timelines and character files, at the height of the controversy, McCloskey had led an aggressive charge to deny Bailey's book a prestigious LGBT literary award for which it had been nominated, and she had helped produce at least one of the formal charges made against Bailey.

Yet in *Crossing: A Memoir*, published in 1999, McCloskey had written the following about Donald, her pretransition self, in the third person:

> When in 1994 he ran across *A Life in High Heels*, an autobiography by Holly Woodlawn, one of Andy Warhol's group, the parts he read and reread and was sexually aroused by were about Woodlawn's living successfully for months at a time as a woman, not her campiness when presenting as a gay genetic man in a dress. Donald's preoccupation with gender crossing showed up in an ugly fact about the pornographic magazines he used. There are two kinds of cross-dressing magazines, those that portray the men in dresses with private parts showing and those that portray them hidden. He could never get

> aroused by the ones with private parts showing. His fantasy was of complete transformation, not a peek-a-boo, leering masculinity. He wanted what he wanted.

An erotic desire for transformation to womanhood? Hello. Reading this passage during my research, I recalled the time I had met McCloskey, well before Bailey's book came out, when she and I were both invited to speak on a panel at her university. McCloskey is a very smart and witty speaker. As I recall, she began her presentation by startling the audience, saying, "These are my cheekbones." She paused while we all sat amazed at her very feminine profile. And then she added, "I paid for them." We laughed at her joke. McCloskey then went on to list other feminine parts she had purchased.

At one point in this anatomical audit, McCloskey talked about how she had had the bone of her forehead surgically shaved back to give her a more feminine head shape. As I remember it, as she explained this, she sort of closed her eyes and talked dreamily about how *thrilled* she had been, the first time she was in the shower and the water ran into her eyes, as it does on a natal woman. First off, I never knew this problem had a sex difference to it. But more important—huh? Why was she saying this as though she was recalling a magnificently sensual moment? Shampoo in your eyes as sexy experience?

And then there was this 1998 e-mail from another trans woman, a letter handed to me during my research by its original recipient, Anne Lawrence, a physician and sex researcher who self-identifies as an autogynephilic trans woman. Writing to praise Lawrence's explication of autogynephilia, the correspondent first acknowledged that many transgender people reject categorization because:

> A definition is inherently inclusive or exclusive, and there's always going to be someone who doesn't feel they belong in or out of a definition. I got body slammed by the usual suspects in 1996 for recommending a Blanchard book. Sure, he's pretty

> much the Antichrist to the surgery-on-demand folks, and I've heard some horror stories about the institute he runs [the Clarke Institute of Psychiatry, in Toronto] that justify the nickname "Jurassic Clarke." However, I found many of his observations to be quite valid, even brilliant, especially in distinguishing early and late-transitioning TS [transsexual] patterns of thought and behavior.

The writer then went on to talk about herself:

> I have noticed in most TSs, and in "surgery addicts" especially, a certain sort of self-loathing, a drive to efface every shred of masculinity. *While I readily admit to my own autogynephilia,* I would contend that my drives towards feminization seem to have a component pushing me from the opposite direction as well [i.e., away from masculinity].

The author of this 1998 letter praising Blanchard's work and readily admitting her own autogynephilia? None other than Andrea James.

OK, THIS WAS FASCINATING. A prior admission to autogynephilia from James and what seemed to amount to the same from McCloskey—plus something very much like an ongoing tacit admission from Conway?—lying behind the attempts to bury Bailey. All that spoke to motivation on the part of Conway et al. Of course, it didn't make them guilty of anything, really (except maybe self-deception). And it certainly didn't exonerate Bailey.

So I dug into this history, never imagining it would end up involving a hundred people and the collection of a few thousand sources, never imagining that it would be like doing a dissertation all over again, only this time with a steady undercurrent of unfamiliar fear. And as I began to dig into this history, it seemed very likely to me that Bailey had, in fact, committed various offenses. I would even have bet on it. There was so much smoke—there just had to be something burning.

But at that point—near the start of the long, unsettling trip—the only thing I felt sure of was this: That now, whenever I found myself standing in the shower trying to keep the shampoo from stinging my eyes, I couldn't help but think hard about what you're supposed to do when the facts seem to be leading you *into* danger.

CHAPTER 3

TANGLED WEBS

THE FORMAL COMPLAINTS as posted on Lynn Conway's site suggested Mike Bailey had dragged a small group of trans women out of the closet and made public spectacles of them, ending their "stealth" lives passing as demure and ordinary women. But I soon learned that the real story was quite different.

A full decade before publication of *The Man Who Would Be Queen*, the trans woman known in the book as Cher—whose real name is now widely known to be Charlotte Anjelica Kieltyka—had sought out Bailey, not the other way around, and Kieltyka had subsequently introduced Bailey to most of the other trans women mentioned in the book. Back in 1993, Kieltyka had seen Bailey on a *Dateline NBC* segment talking about tomboys. Soon after, she called Bailey's Northwestern office, eager to tell him all about herself. Kieltyka wanted Bailey to understand that, despite the media stereotype of transsexual women as extremely femme and sexually attracted to males, she had had a more masculine-typical tomboy history and was attracted to women.

Kieltyka's description made Bailey suspect that her story represented a classic case of autogynephilia, and as if to confirm his suspicions, at their first in-person meeting at Bailey's office, Kieltyka brought as "show and tell" the female masks and prosthetic vulvas she had used pretransition to make erotic films in which she had played a woman. She soon also shared the video with Bailey, including clips that showed a pretransition Kieltyka fully costumed as a female, complete with glue-on vulva, fake breasts, and an elaborate female mask. The film culminated with Kieltyka as a woman simulating dildo-vaginal intercourse. For her part, Kieltyka didn't see all this as evidence of autogynephilia; she saw her pretransition cross-dressing episodes as rituals or "dress rehearsals" leading her to understand that she was really a lesbian woman inside. She believed that her sexual use of women's "foundational garments" had helped her to understand the feminine foundation of herself. But for his part, Bailey saw all this as evidence of *amour de soi en femme*—autogynephilia.

Although he thought her autogynephilic right from the start of their association, Bailey never gave Kieltyka any reason to think he thought *less* of her for having that sexual orientation. Bailey didn't think anyone should be judged for her or his sexual orientation, because he believed that none of us chooses her or his sexual attractions. (He reserved negative judgment for sexual *actions* that directly involved someone who had not consented or could not really consent, like a child.) Indeed, Bailey always found himself *admiring* anyone who admitted to a socially shunned sexual orientation; he saw it as a sign of self-awareness, bravery, honesty, and integrity. As a consequence, he saw Kieltyka's openness and pride about her autogynephilic sexual life history as nothing but admirable, and he let her know it by supporting her desire to present her interesting sexual history to others. For example, he invited her to lecture to the Northwestern students in his Human Sexuality class as part of a series of optional after-class sessions in which students could meet people with the kinds of sexual histories they were learning about in class. Bailey always let his after-class presenters have full control—to say and show just about whatever they wanted, and Kieltyka took full advantage of the opportunity to give elaborate multimedia autobiographical presentations. Twice she even opted to end her appearance by stripping naked. (She said she did this to make the point that transsexual women could be extremely attractive, even in the nude.) Kieltyka's openness with Bailey and Bailey's students—who over the years numbered in the thousands—was not atypical for her; Kieltyka also sought out (and took) opportunities to give public presentations about her life around the Chicago area, including on local television.

About three years into their acquaintance, Kieltyka came to Bailey to ask a favor. Since well before she and Bailey had met, Kieltyka had been acting as a kind of den mother to a sizable group of young transgender women in Chicago, mostly young Hispanic people transitioning from lives as ultrafemme gay men to straight women. Kieltyka (who is white and non-Hispanic) had been doing what she could to help these young people get safe access to hormones and surgery. Around 1996, Kieltyka came to Bailey to ask if he would be willing to write letters supporting her younger friends' requests for sex reassignment surgery. At that time, most of the surgical establishment required letters from two psychological professionals before undertaking surgery on an adult trans patient, an onerous requirement for people without a lot of resources. Again in keeping with his sexual libertarianism, Bailey thought if he could help these capable, adult transgender women get what they wanted out of life simply by having a couple of short conversations and then writing what amounted to a letter of recommendation, he should. It would be up to the surgeon whether the candidate was ultimately accepted, but Bailey's letters might help avoid frustrating expense and delay that these women and he saw as unnecessary. He ended up writing between five and ten of these short letters, including one for "Juanita," the woman who would later claim he had had sex with her when she was his research subject.

In spite of what some critics of *The Man Who Would Be Queen* would later suggest, Kieltyka and Juanita were publicly out as trans women and out about their sexual histories well before Bailey's book. Besides presenting themselves and their autobiographies to thousands of undergraduates, Kieltyka and Juanita also provided their stories for a

1999 article in the Northwestern student newspaper and a 2002 human sexuality educational video. For the newspaper article, Kieltyka and Juanita gave the reporter their life stories, their real full names before and after transition, and photos of themselves before and after. For the human sexuality educational videos, recorded to accompany a textbook Bailey was helping with, Kieltyka and Juanita opted yet again to proudly show their faces, give their real first names, and tell their sexual life stories. In her segment, Kieltyka again showed off her pretransition cross-dressing "props." In her segment, Juanita—the woman who a year or so later would anonymously play a wounded, innocent shy girl outed and sexually used by the ruthless cad Bailey—went on like this, with a confident smile: "When I was a she-male [and] I prostituted myself, . . . I enjoyed it . . . easily making about a hundred thousand [dollars] a year."

Over the years, Kieltyka *did* keep trying to convince Bailey of her vision of herself as something other than autogynephilic in sexual orientation. But the more she talked, the more she just seemed to embody Blanchard's description of autogynephilia. When, near publication, Bailey showed Kieltyka the draft of what he had written about her in his book, she took issue with none of the details about her sexual history, objecting again only to the label of autogynephilic. Understanding she didn't like the label, Bailey and Kieltyka finally decided Kieltyka should be given a pseudonym for the book. She was given the name Cher, similar to Kieltyka's chosen first name (Charlotte) and also the name of the musician whom Kieltyka resembled. Bailey also gave pseudonyms to the other real-life characters in the book, including "Juanita."

It might seem odd that Kieltyka would continue to collaborate and associate with Bailey when she didn't like how he was labeling her. But keep in mind that she heard nothing from Bailey that indicated he felt anything other than full acceptance (and even admiration) of who she was and how she had gotten there. What he wrote about her in the book was what he felt:

> I think about what an unusual life she has led, and what an unusual person she is. How difficult it must have been for her to figure out her sexuality and what she wanted to do with it. I think about all the barriers she broke, and all the meanness she must still contend with. Despite this, she is still out there giving her friends advice and comfort, and trying to find love. And I think that in her own way, Cher is a star.

As he worked on his book manuscript, Bailey's generally warm and appreciative attitude toward all the trans women in his book must have led Kieltyka and Juanita to believe—and no doubt Bailey must have also believed—that his book would help advance the full acceptance of these women. Although he knew of the hatred some other transgender women had expressed online for Blanchard's articulation of autogynephilia, it simply never occurred to him that any of the anti-Blanchard crowd could turn his friends Kieltyka and Juanita into weapons to be used against him. Nothing in their mutual history indicated that that possibility lay ahead.

WHEN *The Man Who Would Be Queen* finally came out, in the spring of 2003, the initial media buzz was positive, and many of the trans women whose stories Bailey had relayed actively helped him promote the book, as they had helped him in his writing. Within a couple of weeks of publication, the *Chronicle of Higher Education* sent its staff reporter Robin Wilson to Illinois to do a feature on Bailey and his book. Wilson, Bailey, and a group of trans women—including Anjelica Kieltyka and Juanita—all went out together to the Circuit nightclub. The resulting article, entitled "Dr. Sex," indicated that Kieltyka had told the reporter Wilson she was not thrilled with Bailey's labeling of her as autogynephilic: "Ms. Kieltyka says the professor twisted her story to suit his theory. 'I was a male with a sexual-identity disorder,' not someone who is living out a sexual fantasy, she says." But the rest of the trans women seemed explicitly and unequivocally supportive of Bailey and his book. Wilson told her readers, "They count Mr. Bailey as their savior." She explained:

> As a psychologist, he has written letters they needed to get sex-reassignment surgery, and he has paid attention to them in ways most people don't. "Not too many people talk about this, but he's bringing it into the light," says Veronica, a 31-year-old transsexual woman from Ecuador who just got married and doesn't want her last name used.

So for the transkids at the outing, including Veronica and Juanita, everything seemed rosy. But for Kieltyka the scene had already started to turn dark. By the time of the get-together with the *Chronicle* reporter at Circuit, Andrea James and others unhappy with Bailey's book had reached Kieltyka to register their displeasure with her "star" turn as proof of autogynephilia. Indeed, it appears that, within days of the book's appearance, Bailey's detractors had figured out who "Cher" was. And they wanted a word.

Now, to the average reader of the *Chronicle,* Wilson's article made it sound as though Kieltyka had been mad at Bailey all along, that perhaps she had been duped into being the autogynephilic subject of his book. But what had changed between the book's publication and the Circuit gathering was not Kieltyka's knowledge of what Bailey thought of her or had written about her. What had changed was that Kieltyka found out she was quickly coming to be considered a pariah by certain transgender activists—the ones who detested any mention of autogynephilia. Kieltyka had found out that

> AJ [Andrea James] and the rest of them wanted to lynch me, as they did Joan Linsenmeier [a colleague and friend of Bailey's who had helped him with the manuscript] and anyone else connected with the book. They were about to hang

> me. I was told this by people that had frequented the Internet, and that's why they gave me the link to contact Andrea James and Lynn Conway, because I was going to be hanged by them.

Yet in spite of this reasonable fear—that she was going to be "lynched"—Bailey and Kieltyka continued to speak with some warmth, each trying to mount defenses against a growing onslaught of criticism. Two weeks after she had read the published book, and one week before the gathering at Circuit with the *Chronicle*'s reporter, Kieltyka had written an e-mail to Bailey using the wry subject line "Cher's Guide to Auto . . . Repair." There she wrote, with her characteristic humor and liberal use of ellipses:

> Dear Mike, . . . I followed up on the links to your difficulties with some hysterical women [an apparent reference to Conway and James] when you wrote . . . "I understand that [trans woman scientist Joan] Roughgarden is slated to review my book for Nature Medicine, and I am certain that this review will be as fair and accurate as her review of my Stanford talk." . . . I really appreciated the sarcasm . . . just wear a bike [athletic] support to your next book signing or lecture. . . . you can borrow mine, I don't use it nor need it anymore. . . . Your friend, in spite of spite, Anjelica, aka Cher

Until things got really hot—until at least a few weeks later—Kieltyka seemed likely to continue her affiliation with Bailey.

As it happened, however, Conway showed up. And not just online, but in person. She started making what she called "field trips" to Chicago. The purpose? "To meet and begin interviewing Bailey's research subjects." Via these field trips and interviews and Conway's and James's Web sites, the public image of the scene quickly changed into "Bailey versus all LGBT folk," such that most people (like me) casually watching the kerfuffle soon thought all the trans women in Bailey's book felt surprised, abused, and angry about the book's contents.

Casual observers thus remained oblivious to something critical: even months into the mess, plenty of gay and transgender people who had read *The Man Who Would Be Queen* actually saw the book not only as an accurate accounting of various forms of "feminine males"—from femme boys to gay men to transkids to drag queens to cross-dressers to fully transitioned autogynephilic trans women—but also as wonderfully supportive of LGBT people. The reason for that would have been obvious, *if you bothered to read the book:* In it Bailey unequivocally supports the right of all people to be gender-variant, to enjoy whatever sexual orientations they have (so long as they're not using anyone who can't consent or hasn't consented), to be recognized by the gender labels they choose for themselves, and to get whatever medical interventions they wish. But most people didn't read the book; they read only the reports of Bailey's alleged abuse. And so they understood this book to be a LGBT-bashing bible—specifically to be to the transgender community what *Birth of a Nation* had been to African Americans.

Nevertheless, if one read the book—something it seems few reporters on the controversy did—one would have quickly realized that it actually made perfect sense that the Lambda Literary Foundation (LLF) included *The Man Who Would Be Queen* as a finalist for the 2004 Lambda Literary Award in the Transgender/GenderQueer category. Although Conway's site would claim that Bailey's publicists had gotten the book nominated for that award, Jim Marks, then executive director of the LLF, later revealed that in fact the book "was added to the list [of nominees] by a member of the finalist committee and after the finalist committee had selected it, we went back to the publisher, who paid the nominating fee." According to Marks, things turned ugly only when, immediately after the announcement of the finalists, Deirdre McCloskey contacted him to express her outrage. McCloskey told Marks the situation "would be like nominating *Mein Kampf* for a literary prize in Jewish studies. I think some apologies and explanations and embarrassment are in order." Marks wasn't sure what to make of all this:

> While I was a little taken aback by the campaign of a university professor to relegate a book to a kind of Orwellian non-history, we might have considered taking administrative action and removing the book from the list if McCloskey's view had been universally that of the transgender community. The LLF was in some senses an advocacy organization. Its stated mission was to advance LGBT rights through furthering LGBT literature. We would clearly have grounds for removing a book that was in fact hostile to the Foundation's mission.

But Marks found that "McCloskey's point of view, although widely shared, was not universally that of the transgender community. Among the torrent of e-mails we received, a minority came from transgender people who supported the book and urged us to keep it on the list." Marks's "main concern was maintaining the integrity of the nominating process." He asked the finalist committee what to do; they revoted, and the vote came back in favor of keeping the book on the list.

A petition sprung up in protest, quickly reaching nearly fifteen hundred signatures. For her part, Conway encouraged her followers to go straight after the LLF committee members. She wrote on her site:

> We thought you'd like to know who the gay men and lesbian feminists are who launched this attack on us. Following are the names, addresses, URL's, and phone numbers of these people. We think that they should hear from you, so as to gain some comprehension of the scale of the pain they have inflicted on trans women throughout the world.

She added a note about lesbian feminist bookstores with a history "of welcoming only 'womyn born womyn'"—a means of excluding trans women—and she "suggest[ed] that our investigators out there quietly gather evidence about any discriminatory policies employed by stores listed below, for future publication on this site."

Meanwhile, under all this pressure, the LLF judges felt the need to vote one more time. According to Marks, on that round, one "member changed their vote and we withdrew the book from consideration." For years afterward, Conway's and James's Web site continued to track Marks, eventually claiming he had been "ousted" over "mishandling of the Bailey matter," something Marks insisted was not true.

The experiences of Kieltyka and Marks were hardly unique. Intimidation tactics flowed in every direction, with Andrea James showing a particular talent for this battle mode. She put up the page about Bailey sodomizing his kids and another page dedicated to making fun of his relationship with his girlfriend, suggesting that Bailey was autogynephilic. Soon anyone who said anything nice about the book became a featured evildoer at James's site. A special circle of hell was reserved for Anne Lawrence, the transgender physician-researcher who dared to describe herself as autogynephilic and to promote Blanchard's work. When in response to threatening e-mails from James, Lawrence refused to back off from support of Bailey and Blanchard, James mounted an extensive attack on Lawrence, making public an incident in which Lawrence had been accused of professional misconduct. James didn't bother to tell visitors to her site that Lawrence had been fully cleared.

James even sent an e-mail message directly to all of Bailey's departmental colleagues, while he was sitting chair, asking why they were allowing "someone suffering from what the DSM calls alcohol abuse and dependence" to lead the department. She told the Northwestern psychology faculty, "I'm sure some of you will continue to respond with self-righteous indignation or with fear of me and my message. For the rest of you, I hope this little rock tossed through your window makes a real human connection." For her part, Conway called Joan Linsenmeier, whom she had found out was Bailey's close friend at Northwestern. Linsenmeier later told me, "I don't recall exactly what she said, but basically it was that some people with very negative feelings toward Mike knew where he lived, that this put him in danger, and that she thought I might encourage him to consider moving."

Of course, most of what was going on remained completely invisible to the outside world. Most people didn't read the book, they didn't know the backstory, and Bailey and his colleagues didn't generally make public the threats to which they were being subjected. (To do so would only have made them subject to more.) Instead, what people on the outside saw were tense news reports about formal charges being filed against Bailey: charges that he had been practicing psychology without a license, that he had been using transgender subjects in research without appropriate ethics oversight, that he had violated confidentiality rules, and that he had had sex with a transsexual research subject. Because Bailey had become worried enough about his job to retain a lawyer, and because the lawyer told him to shut up, the press could not even report what "he said" in response to what "she said." Nothing makes you look guilty like "no comment." As a result of all this, Bailey came across pretty clearly as an abuser, a trans-basher, and a sexual pervert. That was the Bailey I had pretty much expected to meet in Boystown.

As I CREATED TIMELINES out of what sources were telling me about the Bailey book controversy, I suddenly realized something jarring: Even amid the powerful disinformation and intimidation campaigns, there had actually been one reporter who *had* had the opportunity to learn and tell a story much closer to the truth near the start of it all. This was Robin Wilson at the *Chronicle of Higher Education*. As her "Dr. Sex" feature had revealed, Wilson had personally witnessed firsthand the warm relationships between Bailey and his trans women friends and book subjects out at the Circuit nightclub that evening in May 2003. She had herself reported that these women saw Bailey as "their savior." She had spoken to Kieltyka, who—while upset over the label of autogynephile—had had a long and collaborative history with Bailey.

Nevertheless, starting just a few weeks after "Dr. Sex," Wilson published in the *Chronicle* a series of three terribly sober dispatches about the complaints being filed against Bailey. Wilson wrote these scandal reports as if she had just come upon the scene with no previous insider knowledge and no insider connections to use to figure out the truth behind this "controversy." When I realized the strange role Wilson had played, I tried asking her and her editor why they hadn't used her before-and-after-scandal positioning to ask deep questions about why Bailey's relationships appeared, at least in public accounts, to have suddenly changed with these women. Wilson's editor sent me back boilerplate: "We stand by the accuracy, and fairness, of Robin's reporting and are not inclined to revisit decisions Robin and her editors made here with regard to what to include or exclude from those stories in 2003." But I was left obsessing about an *if*: If Wilson had used her special journalistic position as someone who was there just before the mushroom cloud, she might have seen—*right away*—what I saw when years later I charted the journey. What appeared to have happened between the generally happy times that were still evident at Circuit and the unhappy charges was that Conway had shown up at the scene of the alleged crimes, angry about Bailey's promotion of Blanchard's taxonomy.

Now, maybe Wilson would have concluded that Conway had just educated all these women into understanding they had been abused. But if she had taken this or any other theory of what had changed the scene so dramatically, and then *bothered to look into the actual charges*, as I was finally doing years later, she might have seen them fall apart one by one. And then she could have reported *that*. Was Wilson a good liberal simply afraid to look as though she was defending a straight, politically incorrect sex researcher against a group of supposedly downtrodden trans women? Had Conway and James scared the crap out of her, as they seemed to scare everybody else? Or was the explanation simpler? Was it just that trying to figure out what the hell was really going on would have taken too much time and other resources?

Well, such an unquestioning approach wasn't good enough for me. The more I dug, the more I wanted to find out the truth about all of the charges made against Bailey. With the distance of several years, I had an advantage: Many sources that might have otherwise been covered up if someone like me had been doing an investigation were right out there in the open, including on Conway's gigantic Web site. And I could ask people to talk to me—people who at the time of the controversy might have been too afraid. Not everybody agreed to talk to me on the record. From the start, Conway refused, as did Juanita, but ultimately the great majority of people I contacted responded. Bailey was willing to answer any question and open his records to me. Kieltyka was similarly forthcoming, although at first I was worried about even talking to her, given her record of filing complaints. I decided to use for her a system I then used for all oral interviews, to protect myself as well as my subjects: When we talked, I wrote down what I heard, but then gave back the source the notes to change however she or he wanted. Sources could add, delete, and append anything they wanted. Only what the person returned counted as on the record.

Boy, what I got on the record! The interviews for this project turned out to be almost as emotionally jarring as the interviews I had once done with intersex people, so passionate were my sources. Indeed, after almost eleven hours of interviews with Kieltyka, I felt about as sorry for her as I was coming to feel for Bailey. Not only did she feel used up and spit out by Bailey, but she felt the same about Conway and company. They had swept into Chicago after the book's publication, encouraging her to file complaints against Bailey, only to later dump her. By the time I talked to her, she had come to a conspiracy theory in which Conway had actually been using Bailey as a fall guy in a much larger anti-LGBT scheme. I couldn't really follow that complex worldview, but her specific answers to my straightforward factual questions helped fill gaps in the written record.

It became steadily clearer that what Bailey had done wrong was both sadder and much less scandalous than we had all been led to believe. What did Bailey do wrong? Well, in retrospect, it's clear he should have let Kieltyka know that it would be almost impossible to ever convince him that she wasn't autogynephilic. If her hope of changing his mind had been what had kept her collaborating with him, ultimately exposing her to criticism nationally—although I wasn't convinced it was what kept her collaborating with him—then he should have let her know in no uncertain terms that he was very unlikely to ever change his mind about her identity. I also found myself wishing he had started working much earlier to protect Kieltyka's identity, so that she could avoid people like Conway who might eventually go after her for putting a human face on autogynephilia.

That said, from everything I could find, it seemed pretty clear that no matter what Bailey said or did, Kieltyka would have kept presenting herself and the details of her sexual history all over the place, with her face and real name. She had outed herself long before his book. The more I thought about it, the more it appeared that not only was Bailey not guilty of outing her, but he also could have justifiably portrayed Kieltyka in his book using her *real* name while identifying her as autogynephilic if he had wished. There were enough highly detailed public self-representations by Kieltyka that he could have simply drawn from those. Was it personally obnoxious of Bailey to label his friend with a term she didn't want? Certainly. But she had hardly been secretive about the details of her sexual history, and he had not hidden his belief that she was autogynephilic. Bailey's position was like having a friend who was obviously a homosexual man—who openly dated and partnered with men—but who got upset if he heard someone refer to him as a gay man. I suppose one ought to try to observe anyone's preference for self-labels, but doing so can feel like playing pointless games. I got why Bailey had very little patience for a situation in which all the identifiers are there but you're not allowed to apply a category label that you (and many with the identity) don't see as inaccurate or offensive.

In spite of the real backstory of generally warm collaboration and openness, Conway had made it look as though Bailey had dragged all these trans women into his book and outed them without their knowledge and consent. One of the ethics charges made formally to Northwestern by three trans women for whom Bailey had written recommendation letters involved the claim that Bailey had treated what he learned in the conversations leading up to those letters as research material for his book. This supposedly represented an intentional breach of confidentiality. Yet Bailey was able to show me that two of those three women were *not even in the book.* (None of the reporters relaying this charge did the math.) The third complainant? Juanita. Bailey denied that he used in his book what he had learned in the short interviews that went into Juanita's recommendation letter. I had no way to verify the truth of his denial, but I *could* verify this: Juanita had told the Northwestern investigatory committee convened in response to her complaints that she had known Bailey was writing about her in his book and that she had given him permission to do so. Moreover, before Bailey ever submitted his book for publication, like Kieltyka, Juanita had chosen to repeatedly tell intimate details of her life history publicly, to the Northwestern reporter (remember: giving her real names and photos before and after transition). She'd done the same thing in the video filmed with her written consent for a human sexuality textbook. And she'd done the same thing in person in Bailey's after-class presentations, where the audience ultimately added up to thousands. I was also able to confirm, from written documentation provided to me by Bailey, that Juanita had met with Bailey at a local coffee shop to help him write up her story for his book.

It sure didn't look to me as though Bailey would have needed to mine those letter-interviews to tell Juanita's story in his book, what with all her cooperation and steady openness about her life history. From everything I could find, before Conway showed up, Juanita seemed downright *excited* about having her life story out there year after year in the public realm. She was like Kieltyka that way. In fact, the only time Juanita had ever had a pseudonym attached to her story was when Bailey had decided, late in his book project, to assign her one.

The record was clear: Of the four trans women personally known to Bailey who filed complaints to Northwestern about Bailey's book, only two were in the book—Kieltyka and Juanita. These two had known Bailey was writing about them and had helped him. They had also known for years that he was writing about one as an autogynephilic transsexual and the other as a homosexual transsexual. Additional evidence came from a "sealed" complaint Juanita sent to Northwestern, posted on Conway's Web site. There, Juanita said, "an early draft was not objectionable, but *absolutely nothing like* the spurious and insulting description he wrote about my life that did become part of that most hurtful book of his." But in fact the only real changes from the "early draft" to the published version were mentions of Juanita's marriage and subsequent divorce. Juanita knew Bailey was writing about her in his book as an example of homosexual transsexualism, and she raised no objections until Conway appeared with her own objections to Bailey's promotion of Blanchard's two-part taxonomy that saw Juanita as a kind of trans woman fundamentally different from Conway.

What about the claim that Bailey was practicing psychology without a license? This complaint was formally filed by Conway, James, and McCloskey with the Illinois Department of Professional Regulation. The trio's claim found its basis in an assumption that Bailey would have needed a license as a clinical psychologist in order to provide letters of recommendation for the young trans women seeking sex reassignment surgery. As with just about every other cleverly packaged complaint, anyone watching casually from the outside would think this complaint was right on: Bailey, a psychologist, had written letters supporting requests for surgery, but didn't have a license to practice as a clinical psychologist. Guilty! Right?

In fact, as the complainants surely would have seen by looking at the letters they used as evidence, Bailey never pretended to have done therapy with the trans women for whom he wrote letters of recommendation. Not only were these letters obviously based on a few short conversations, they specifically explained Bailey's credentials, and therefore indicated the limits of those credentials. If it wasn't clear to a surgeon from the letter that Bailey wasn't a therapist engaged in deep identity analysis with these transitioning women or anyone else, the curriculum vitae Bailey attached to each letter would have driven home the point. Most important, the relevant Illinois regulations state that if an individual doesn't get paid for services offered or rendered, that person is not required to have a license *even if* he or she is offering what looks like "clinical psychological services." Bailey never took a cent from these women. Presumably all this was why the Illinois department never bothered to pursue the charge, although you'd never know that from reading the press accounts, which mentioned only the complaints, not that they had petered out.

Considering what had really happened, I had to conclude that this claim about practicing without a license wasn't just false. *It was appalling.* Conway, James, and McCloskey had tried to use Bailey's letters of support for trans women to string him up, whereas in providing those letters without great delay and without extracting thousands of dollars in therapy charges, Bailey had been trying to help lower barriers to wanted interventions for these women, *exactly* as many trans activists had sought for their community for years! The letters should have been cause for trans advocates to praise Bailey, not bury him.

Conway and company had also accused Bailey of violating federal regulations. They broadcast the claim that in "researching" trans women for his book, he had been conducting human-subjects research of the sort that requires approval and oversight from a university's institutional review board, or IRB. They said he hadn't got the required IRB approval. Again, to people outside academia, this sounded like a slam dunk. He'd written up these women's stories in a book about "the science of gender-bending and transsexualism" (as the subtitle said), so surely they were "human subjects of research," right?

No, actually not—and what really troubled me is that at least McCloskey (a wizard of language and categories) surely should have known this. IRB regulations, which were originally designed for invasive biomedical research, count as human subjects only those individuals who are enrolled in research that constitutes "a systematic investigation, including research development, testing and evaluation, designed to develop or contribute to generalizable knowledge." For the purposes of his book, Bailey wasn't engaging in novel scientific research of this type; he was just picking and choosing stories from real-life people he met to illustrate scientific theories he believed were already firmly established. One might try to claim (as complaints against him hinted) that in choosing whom to write about in his book, Bailey was engaging in psychological research to test Blanchard's theory. But that would attribute to Bailey a more open mind than he in fact had about male-to-female transsexualism. The truth was that he had become a convert to Blanchard's taxonomy long before he wrote about it. To say Bailey had been doing novel science in his book would be like saying that if you were on a walk with an evolutionary biologist and she chose to point out to you an evolutionarily interesting behavior of some nearby birds, she was doing research to test the theory of evolution. The personal stories in Bailey's book were really just window dressing for a store Bailey had long since bought.

Now, Bailey *had* in fact enrolled some of the trans women he'd met in formalized scientific studies. This occurred, for example, when his lab was studying sexual arousal patterns in adult humans. The lab measured genital blood flow (something sex researchers believe indicates arousal) in natal men, natal women, and trans women while they were shown various kinds of pornography. But was Bailey studying the arousal patterns of these trans women in his lab without IRB approval? Nah. He had full approval for those laboratory studies. And of course he would have, because he understood that that was science, but telling stories about people to bring to life various theories is not.

What about the sex charge made by Juanita? When we got to discussing this, Bailey insisted, in his usual sexual-libertarian style, on discussing the principle at stake. He pointed out that there are plenty of instances where we might not find it unethical for someone to have sex with a person who happened to be a subject in his or her research project, particularly if the research subject were a competent adult, if the sex were unlikely to compromise the research, and if the research subject otherwise had no complicating relation to the researcher (e.g., was not also a patient of the researcher). I found his argument initially startling but ultimately convincing. The more I thought about it, the more I realized there could be instances where the subject, the researcher, and the research would stand no particular risk if sexual relations occurred. Consider, for example, a five-year study of cholesterol levels in which an involved researcher and subject end up in their private lives having sex with each other a few times. Or imagine an anthropologist who, after living with a study group for years, ends up marrying and having sex with someone in the study group and continues the research for several years more.

Now, if a researcher were also a clinician treating a subject as a patient, that would make sex verboten. But a research relationship is not (and should not be) anything like a therapeutic relationship. Often research relationships are brief, impersonal, and unlikely to be compromised by sex. It seems silly to treat all research subjects (or all research projects) as so fragile that they will necessarily be put at risk by sexual relations. Grown adults in a research relationship—capable of having the consenting, mutually respectful relationship required by research—could in some cases legitimately decide that sexual relationships would not entail harm to them or the research.

Still, I wanted to know, *did Bailey and Juanita have sex*, as she said? When I looked into this charge, I was surprised to discover that the notarized affidavit making this claim—posted on Conway's site—consisted of only two sentences: "On March 22, 1998, Northwestern University Professor J. Michael Bailey had sexual relations with the undersigned transsexual research subject. I am coming forward after I learned he divulged his research findings about me in *The Man Who Would Be Queen*."

This affidavit would have us believe that Juanita decided to come forward with this charge *after* (i.e., as a result of) finding out what Bailey had written about her. But the timing simply could not be true. The affidavit was dated July 21, 2003—*months after* Juanita had gone to Circuit to help promote the book, *months after* she had seen a draft, and *four years after* the student newspaper article describing Juanita's role in Bailey's book. These facts clearly contradicted the affidavit's claim about order of events, a claim no doubt made to explain why Juanita waited five years to make the complaint. In addition, in spite of what the affidavit said, Juanita was not a research subject of Bailey by any normal definition on March 22, 1998. This claim would have been true only if the definition of "research subject" had been so broad as to include everyone you ever write about.

That said, I admit I was still dying to know whether they had had sex. Looking at photos and videos of Juanita, including an erotic seminude photo of Juanita that Conway posted on her site, I could well imagine that a straight guy who is not transphobic would be interested. (She's gorgeous.) Alas, Bailey managed to ruin even that possibility. Knowing he had said previously in a rare public statement that he could prove they hadn't done what she said if he needed to prove it, I pressed him for the proof. He promptly showed me documentary evidence that he was home with his kids in Evanston the night Juanita had claimed that she and Bailey had been getting it on in Chicago. His ex-wife had been away on her annual spring break, and by their usual agreement, Bailey had been at her home taking care of the kids and, based on the e-mail reminders from his ex-wife that Bailey showed me as his proof, also taking care of the children's fish, hamster, and cat.

When I asked Mike's ex-wife, Deb Bailey, to confirm the dates and arrangement for me, she did. Tellingly, she was happy to help. Contrary to Andrea James's portrayal of the family as pathological and dysfunctional, the Baileys remained close friends post-divorce, sharing parenting duties, as well as meals at holidays—and also sharing a unified defense against Andrea James.

The more I looked into the sex charge, the more it looked like a setup. Not only had the affidavit been witnessed by Andrea James and Lynn Conway, but the letter accompanying it to Northwestern specifically credited "Lynn Conway and Deirdre McCloskey, who have acted on our behalf to make Dr. Bailey accountable for his actions." Nevertheless, I pressed Bailey. Might Juanita just have gotten the date wrong? Was he using a Clintonian definition of the phrase *sexual relations*? He was adamant, saying that although he had flirted with Juanita once or twice when they were socializing, there had never been anything that could be construed as sexual relations.

Unable to reach Juanita with an interview request through other channels, I asked Kieltyka to put me in touch with her. Kieltyka responded that Juanita wasn't interested in talking to me. But Kieltyka herself was willing to elaborate on the sex charge:

> [Juanita] told me the day after Bailey drove her home from the Shelter nightclub that Bailey had tried to do something. . . . That they had "messed around"—She was being slightly evasive and uneasy so I left it alone. [Five years later, in the summer of 2003] when Lynn Conway [was] over my house, Juanita was there, and that's when she told the two of us that Bailey in fact had had sex with her. This was the first time that I found out it wasn't that he had 'tried something'—it was that he had tried to have sex with her. But that he couldn't get it up.

Wait, Bailey had tried something but *failed*? I pressed on.

DREGER: So you're saying she said he tried but he didn't get it up?
KIELTYKA: Right.
DREGER: And she told that to Conway and McCloskey.
KIELTYKA: Right.
DREGER: And then [in the formal charge] to Northwestern she said that they had had sex.
KIELTYKA: I'm not sure what the letter says. . . . I think it says "sexual relations"—just like El Presidente Clinton. . . . It all is a matter of a definition of what sexual relations is. Because there was fingering, that she was giving him a hand job, I don't recall exactly.

Kieltyka seemed to explain it all when she ended the conversation this way: "Anyway . . . from the moment that Andrea James and Conway wanted to use the sex with a research subject as a way of getting Bailey, I wasn't enthusiastic."

AFTER NEARLY A YEAR of research, I could come to only one conclusion: The whole thing was a sham. Bailey's sworn enemies had used every clever trick in the book—juxtaposing events in misleading ways, ignoring contrary evidence, working the rhetoric, and using anonymity whenever convenient, to make it look as though virtually every trans woman represented in Bailey's book had felt abused by him and had filed a charge.

"Narcissistic injury," the physician-researcher Anne Lawrence said to me, by way of explanation. "Followed by narcissistic rage." That, she told me, was the only real way to explain what happened to Bailey. The whole thing had been an attempt to kill the messenger bringing a message that Lawrence guessed wounded the accusers' senses of self. They didn't want to hear what Bailey said, so they had to make him just go away—and make sure no one else ever tried it again.

For the sin of speaking honestly about autogynephilia, Anne Lawrence had become the third leg of the "Bailey-Blanchard-Lawrence Axis of Evil." Yet somehow, throughout the attacks from Lynn Conway and Andrea James (who had once been her friend), Anne had soldiered on, publishing narratives from trans women like herself who are autogynephilic, doing research that showed that autogynephilic males who want sex-reassignment surgery and are screened by professionals are on average better off after surgery, and providing clinical care to those who needed to be reassured, from a woman who knows, that being autogynephilic in your sexual orientation doesn't make you less genuinely transgender, no matter what some self-appointed "trans advocates" say. Anne had even worked to change the official standard of care so as to provide easier access to sex-changing hormones and surgeries for all trans people. When we became friends, she showed me a photo of her, Mike Bailey, and Ray Blanchard sitting on a couch together and doing that little pinky-on-the-lips thing that the villain does in the Austin Powers movies. It cracked me up.

But God almighty, what this crew hadn't been through. And not just them. James had also gone after other self-identified autogynephilic trans women, and also after self-identified transkids (the group Blanchard had called homosexual transsexuals). About those attacks, one day Paul wrote to me to say that there was a Bailey-defense Web site, www.transkids.us, edited by Kiira Triea. Could that be the same Kiira Triea of the intersex rights movement, he wondered, the one who had made the satirical phall-o-meter and had provided her autobiography for my edited collection, *Intersex in the Age of Ethics,* now wrapped up in the Bailey book controversy? I looked at the site. How many Kiira Triea's with that wicked a sense of humor could there be?

Thanks to Paul's tip, I reconnected with Kiira, whom I hadn't talked to since we worked on the intersex book in 1999. In that book, she had told readers that her mother had been given progestin to prevent a miscarriage when she was pregnant with Kiira. Although Kiira was genetically female and had ovaries and a uterus, her genitals looked male when she was born because of the progestin. So Kiira was labeled and raised as a boy, but turned out to be a real sissy boy, and by puberty a boy-crazy femme boy. Eventually recognized as intersex, Kiira landed in John Money's clinic. Money was deeply annoyed at this fourteen-year-old who was messing with his theory that gender and sexual orientation result from nurture, not nature. Money tried to use masculinizing hormones to make this annoying boy more boyish. When they failed, Money "let" Kiira become a girl and gave her sex reassignment surgery. So Kiira was not just intersex but had had the life of a typical transkid, too. She'd been a femme gay boy who became a woman. No wonder she related so easily to other transkids. No wonder she was defending Bailey.

Kiira gave me all sorts of much-needed sympathy over my run-in with Andrea James (she'd had one too) as well as over my unwanted exit from ISNA (she'd had a similar thing with Bo). Honestly, it was like one of those weird moments when you're in Paris and you take a wrong turn and find yourself in some back alley and run smack dab into a friend you haven't seen since college. I asked Kiira to help me understand it all. Basically, she explained to me, Bailey and Blanchard and Lawrence were right. She pointed out that Mike had been incredibly thickheaded about some things, like about his assumption that transkids end up prostituting themselves and shoplifting because that's what they like to do. That's how they survive, she said, adding an observation from her transkid friend Alex: "Claiming that transkids may be especially suited for prostitution because they sell themselves when they are destitute is like believing that people in famine ravaged countries are especially adept at dieting."

That aside, Kiira said, Blanchard's taxonomy was right. She told me to think back to all those trans women who joined ISNA, wanting to be told they were hermaphrodites, looking for some explanation for their sexual feelings and their feelings that they were women in spite of a life of masculine signs. Kiira reminded me how decent many of those trans women had been, helping us early in the intersex-rights movement. I immediately thought of Maxine Petersen, a post-transition self-identified autogynephile who had worked in Ray Blanchard's clinic, helping women like her. I thought of four other trans women friends whose friendship I now knew I could never dare to mention in public, lest they be taken as hostages in this war. Those are the people who need you to tell the truth about what happened, Kiira said to me, meaning the transgender people whose identities were being disappeared by those bent on a self-serving narrative.

She was right, of course. As scared as I was getting, I knew I had to write up what I had found out about the Bailey book controversy. Even while my stomach hurt from the thought of the backlash that would surely be directed at me, the scholar and the activist in me felt as I had at the start of the intersex-rights movement: that I was suddenly seeing a truly oppressed population who had been made nearly invisible by people inconvenienced by their reality. Only in this case, it had happened at the hands of their own kind.

THERE ARE NOT a lot of places that will take a book-length academic article, even one that people consistently call "a real page-turner." I negotiated consideration of my tome by the *Archives of Sexual Behavior*, the highly rated journal of the old-boy sexologists. It was true that Mike, Paul, Anne Lawrence, and Ray Blanchard were all on the editorial board, and that would look funny. But it was also true, I knew, that no one could ever really dispute anything important in what I had found. Plus, surely nowhere else would take this kind of article under the terms I wanted: I wanted the right to make it available to all for free online, and to have it published on paper as well, full-length, uncut, so that it would be readily accessible in all reputable academic libraries. I also insisted that Conway and Bailey both be given an opportunity to respond in the very same issue in which my work was published. The editor of *Archives*, the child psychologist Ken Zucker, told me he'd go one better and let *anyone* write a response and publish anything reasonably related to my article. Up to that moment, Zucker and I had only had tense encounters, as my work on intersex had steadily criticized the pediatric gang with whom he ran.

Using this open-dialogue approach to my "target article" meant Zucker had to release a copy of it after it was peer-reviewed but before it was published, so that people could write their responses for the dedicated issue. I knew from my intersex experience that if I wanted to get anywhere in terms of public understanding of what had happened, I needed the press to get in on the story early. So I sent out a cold-call e-mail to Benedict Carey of the *New York Times*, and to my relief, he replied with interest. Carey had previously written about Bailey's work on bisexuality and had taken crap from none other than Conway for it. (She went after Bailey on everything.) This prior history made me nervous about appearances: Didn't Conway's previous attack on Carey give him an apparent conflict of interest, I asked? He explained to me that if people were allowed to use any criticism to neutralize reporters, the free press would die, and his editors understood that. I felt a glimmer of hope for the Fourth Estate.

With a copy of my paper in hand, Carey did his homework, calling all the major characters, checking my claims against theirs. Conway refused to talk again, but it was now obvious that they all knew what I had found and that they were worried, because they started to really come after me online. As I waited after the release of my article, wondering if the *Times* would ever report on it, I watched my Google rankings being taken over, almost in an instant, by James and Conway's claims about me. I was being reconstructed as an enemy of intersex people and an enemy of all LGBT people. And how did they portray my exit from Michigan State? Predictably, my choice to give up tenure was simply reconstituted as my having been driven out for bad behavior.

Finally, Carey's piece was published in the *Times*, and he amazed me by his ability to sum up the salient points in a couple thousand words. More important, Carey's report turned around the public story of what had really happened. Mike was elated. Mike's family was elated. Ray Blanchard was elated. Scientists all over the world were elated.

Me—well, it's not as though I didn't know what was coming, as though I didn't know that it would just get worse after the positive press coverage of my work, first in the *Times*, then in the *International Herald Tribune*, then even in the *Advocate*, the national magazine dedicated to LGBT politics. But somehow it felt shocking anyway—page upon page on the Web, "exposing" me as a right-winger, a fake, a eugenicist.

For her part, McCloskey wrote to the *New York Times* to say that "the Bailey group" had paid for my work, and she put a copy of her letter to the editor on her Web site. The truth was that I had paid for the whole thing out of my own puny part-time income, and that I had felt the need to try to hide the project from Northwestern as much and as long as possible, worried that they'd fire me for stirring up more trouble from this crowd. It would have been very easy for them to do so; I was on a one-year, part-time job with no contract. Several of my colleagues had made perfectly clear that this one "wasn't worth it." I wasn't sure Northwestern would now feel *I* was worth it. I tried to remind myself I'd been looking for a way out of academia. But I also tried to get Northwestern's general counsel to tell McCloskey to stop defaming me by saying Bailey had paid for my work. They told me they didn't work for faculty; I'd have to get my own lawyer.

Honestly, I have forgotten a lot of what happened during the worst of the storm. I remember debating Joan Roughgarden of Stanford (the trans woman scientist) live on public radio in the Bay Area, and getting really upset at listening to her not just repeat the false charges against Bailey, but also inflate them. I remember having to put my friends under strict rules not to tell me what they were finding on the Web about me. I remember being afraid to open my e-mail, even though so much of it came from trans women thanking me for "telling the truth." I remember trying really hard to focus on the fact that I had a PhD and that no matter what they did, no one could take that from me. I remember hosts of my invited academic lectures calling to tell me they were getting angry mail about my having been invited to speak and asking me if they needed to call campus security. I remember one morning, sitting on the kitchen floor and crying into my hands as quietly as I possibly could, while my son sat in the next room waiting for his breakfast.

I think the lowest point was one Saturday night when I opened my mail near midnight and found out that a trans woman named Robin Mathy was filing ethics charges against me with my dean. Mathy considered me unethical for publishing the work in *Archives* (because it was the journal of the sexologists, including Bailey, Blanchard, and Lawrence), and unethical for concluding, in my article, that sex with a research subject is not always problematic. In that bit of my article, I'd gotten personal, saying that if everyone we write about is a research subject, and sex with research subjects is always wrong, then I had violated that rule because I often write about my husband. I mean, once you understood what Mathy was really charging me with, her claim was ridiculous—especially given that she also made the complaint to the American Psychological Association, and I'm a historian. (I felt like suggesting she also send her complaint to the American Dental Association and the greengrocers of America.) But she had professional mental health credentials and a trans identity card, and she was good at making things sound official, and I knew perfectly well that now they would all use this to say, "Dreger has been charged with ethics violations involving a question of sex with research subjects." And people wouldn't look up the details. They never look up the details.

Perhaps most disorienting of all was now being held up as a hero by all these guys whose work I had criticized. Even Steven Pinker from Harvard—who had gotten swept in years ago by blurbing Bailey's book—wrote to say how impressed he was with my article. He offered me an introduction to his agent. In reply, I asked him to support an application for a Guggenheim Fellowship. I needed to feel that I was still a scholar, no matter what they said about me.

In my application, I proposed to look at conflicts involving scientists and activists over matters of human identity as they play out in the Internet Age. What's that the Brits say? "In for a penny, in for a pound."

CHAPTER 4

A SHOW-ME STATE OF MIND

IT WASN'T LONG before I had lots of good company for my misery. As soon as word got around that I had a Guggenheim Fellowship to study conflicts between scientists and activists over issues of human identity, academics from all over started contacting me, suggesting I work on this controversy or that. I heard from one physician colleague about a clinician-researcher who dared to question the reality of chronic Lyme disease and was now chronically plagued by people who insisted they had it. I heard from another about the physician-researcher who had helped to define the condition known as fibromyalgia only to later doubt that it really is a distinct disease. (There's a way to make yourself researcher non grata.) I started to wonder if this was just a guy thing. Are men much more likely to get into trouble because they're taught and allowed to be aggressive? Then Mike Bailey told me about another woman who'd been in this kind of trouble, a clinical psychologist who had researched and revealed the disappointing reality behind a poster-child case of "recovered memory" of alleged childhood sexual abuse. Then I learned from an editor at Harvard University Press about another woman psychologist, one who had experienced some significant unpleasantness following a book in which she expressed scientific skepticism about alleged alien abductions. The abductees wanted a word with her.

I had accidentally stumbled onto something much more surreal—a whole fraternity of beleaguered and bandaged academics who had produced scholarship offensive to one identity group or another and who had consequently been the subject of various forms of shout-downs. Only these academics hadn't yet formed a proper society in which they could keep each other company. Most of these people had been too specialized or too geeky (or too convinced *they* were the only ones who didn't deserve it) to realize there were others like them out there. As I started collections of notes on each of these folks, I kept thinking about how Bo must have felt in the early 1990s, when she realized that there were others like her out there, others who had been born with ambiguous sex, who had been cut, changed, and lied to. "My people" were out there. I just had to put it—put them—all together.

But where to start? Mike Bailey's son, Drew, insisted the place was the University of Missouri in Columbia. Drew was there earning his PhD in evolutionary psychology, a field I had long held in contempt as I knew feminist science-studies scholars like me were supposed to. Drew assured me I could get plenty of material from a single trip to the University of Missouri, because the place was littered with CV's torn asunder in various controversies. There was Craig Palmer, the anthropologist who had dared to cowrite *A Natural History of Rape: Biological Bases of Sexual Coercion.* There was Ken Sher, the psychologist who had been the action editor for an infamous paper purporting to show that children are not, on average, as universally devastated by sexual abuse as the angriest survivors might lead us to believe. There was Dave Geary, a psychologist who dabbled dangerously into the study of sex differences in mathematical abilities (even after Larry Summers). And there was Mark Flinn, a scientist whose career had been wrapped up with Napoleon Chagnon, the famous sociobiology-loving anthropologist who had been tried for high crimes and misdemeanors by the American Anthropology Association in 2001.

I wrote to all these people and set up interviews for the two days I would spend in Columbia in late October 2008. Then I crammed, studying these various people's experiences, and got on a plane from Michigan to Memphis, because (as I had learned) Memphis was the only way to fly to Columbia. Turns out Columbia sports a tiny airport with a total of three flights in and three out each day, except on Saturdays when they drop down to two each.

As we approached the landing strip in the midst of green rolling hills dotted with brightly colored trees, I suddenly wondered something. If in a place as small as Columbia, I could quickly find people with intimate connections to at least four major controversies involving scientific claims about human identity, how many of us must there be?

OBAMA SIGNS WERE EVERYWHERE as I pulled my rental car into Columbia and found a place to park. It was five days before the presidential election of 2008, and Missouri had become a swing state. I knew that unless the polls suddenly shifted, Obama would be in Columbia the next night for a rally. Everyone seemed downright giddy, like the mood of the home-team fans as the clock ticks down on a national championship, the scoreboard shining with a point lead clearly too great for the visitors to overcome. The academics in Missouri were as academics everywhere that year—gaga over Obama not just because he could really think and reason and write and speak, but also because the Bush administration had been so profoundly antiscience and so very untruthful. Such *liars.* We just wanted reality back.

With all this going on and not really knowing if I was at the beginning or near the end of this shape-shifting project, it was impossible not to feel disoriented. I was glad when Ken Sher, my first interviewee, suggested we talk in a bar. It was hard to find a place to perch my computer for note taking, but I didn't really care. The controversy he'd been involved with had already been exceedingly well documented, including by people like Ken. I mostly wanted to hear what Ken thought I needed to really appreciate about how these things play out.

I knew from my background reading that in 1998 Nancy Eisenberg (of Arizona State) and Ken had been the editors at *Psychological Bulletin* responsible for publishing a paper that came to be known as "the Rind paper." *Psychological Bulletin* is one of the publications of the American Psychological Association, and at the time, Eisenberg served as editor in chief. As was typical for manuscripts submitted to the journal, Eisenberg assigned the submitted Rind paper to one of her two associate editors to shepherd it through peer review and revision. The associate editor assigned the task of managing the manuscript was Ken Sher.

Sher and Eisenberg had decided that the Rind paper—after it passed the usual peer-review process—would certainly be worth publishing. Bruce Rind, Philip Tromovitch, and Robert Bauserman had performed a meta-analysis of studies of childhood sexual abuse, or CSA. They took a series of existing studies—on college students who as children had been targets of sexual advances by adults—and looked to see what patterns they could find.

What had seemed particularly important to Sher and Eisenberg about the Rind paper was its parsing of which factors in cases of CSA were associated with long-term psychological harm. For example, the Rind paper's analysis suggested that girls were more likely to be psychologically harmed by CSA than boys and that an incestuous family environment that also involved other forms of physical and emotional abuse was more likely to result in harm than nonincestuous CSA in less abusive environments. The paper also suggested that it did not make a lot of sense to lump together under the term childhood sexual abuse both (a) molestation of a five-year-old by a sixty-year-old, and (b) consensual sex that happens between a sixteen-year-old and a twenty-year-old. Yet the scientific literature sometimes did that. The paper therefore tried to bring a more scientific approach to a very heated topic.

But in taking seriously the idea that not everyone is devastated by everything termed childhood sexual abuse, Rind, Tromovitch, and Bauserman were saying something very politically incorrect: some people grow up to be psychologically pretty healthy even after having been CSA victims. In fact, Rind and company opted to go even further in their paper, suggesting that the term *childhood sexual abuse* seemed to imply that child-adult sex always led to great and lasting harm, whereas the data seemed to show it did not in a surprising proportion of cases. The Rind paper recommended that those studying the problem employ a more neutral terminology and sort out the different types of sex (and harm) occurring.

Rind, Tromovitch, and Bauserman closed the paper with an attempt to avoid being accused of being apologists for pedophilia, reminding readers that just because an action might not harm does not make it morally right: "If it is true that wrongfulness in sexual matters does not imply harmfulness"—a point they attributed to my old pal John Money—"then it is also true that lack of harmfulness does not imply lack of wrongfulness. . . . In this sense, the findings of the current review do not imply that moral or legal definitions of or views on behaviors currently classified as CSA should be abandoned or even altered."

Well, that little "we're not advocating pedophilia" disclaimer sure didn't work. Activist pedophiles saw in the Rind paper justification for us all to just get out of the way already and let them at little kids. Blatantly ignoring the point made at the end of the Rind paper, the North American Man/Boy Love Association (NAMbLA) called the Rind paper "The Good News About Man/Boy Love." Thus the Rind paper became the gospel according to NAMbLA, a group whose mere logo can thoroughly creep you out. (The *M* for Man leans to the right, pushing against the little *b* for boy, as though the *M* is mounting the *b*. Seriously, that's what it looks like.)

If NAMbLA saw a golden opportunity in the Rind paper, Laura Schlessinger had visions of platinum. In the spring of 1999, on her *Dr. Laura Program*, Schlessinger simplified the whole scene in predictable ways, making the Rind paper out to be junk science and suggesting that Rind and company were virtual pitchmen for pedophilia. The fact that the American Psychological Association (*Psychological Bulletin*'s publisher) then looked like the PR arm of NAMbLA was undoubtedly a delightful side effect from Schlessinger's point of view. Schlessinger had no use for the APA, an organization that openly leaned left. She pulled out all the stops, ultimately encouraging conservative members of Congress to use the Rind paper to go after the APA. The not so honorable Tom DeLay, representative of Texas's Twenty-Second Congressional District, was only too happy to heed Dr. Laura's call. No doubt still stinging from impeaching Clinton without managing to remove him as president, DeLay found in the Rind paper a new sexual ticket to ride. And ride it he did.

By July of 1999, DeLay managed to get the House of Representatives to condemn the Rind paper by a vote of 355 to 0 (with 13 members voting only "present"). Just a few days later, the Senate followed suit, unanimously resolving "that Congress condemns and denounces all suggestions in the article 'A Meta-Analytic Examination of Assumed Properties of Child Sexual Abuse Using College Samples' that indicate that sexual relationships between adults and 'willing' children are less harmful than believed." Congress also saw fit to condemn "any suggestion that sexual relations between children and adults . . . are anything but abusive, destructive, exploitative, reprehensible, and punishable by law."

And there you have it: the only scientific paper ever to be condemned by an act of Congress. Sher and Eisenberg found themselves getting mail like this:

> WHO THE HECK CAME UP WITH THE IDEA THAT BEING RAPED IS OKAY IF YOU HAPPEN TO BE A LITTLE CHILD? WERE ANY OF YOU VICTUMS [*sic*] OF SEX ABUSE? BELEIVE [*sic*] ME YOU WOULD BE CALLING IT ABUSE IF IT HAD HAPPEND [*sic*] TO YOU. . . . SO BEING SEXUALLY ABUSE [*sic*] MAY NOT BE A

> CRIME TO YOU BUT IT IS TO ME. . . . CUT OUT THIS INSANITY AND JUST SAY THAT YOU ARE SORRY AND ARE IN ERROR.

Sitting at the Missouri bar with me nearly a decade later, Sher seemed to alternate between a cringe and a sardonic smile as he recounted to me the insanity of it all. I mean, how do you even start to explain what the Rind paper actually said when you're dealing with Dr. Laura's distorted caricature? And as for DeLay—it's one thing to use your elected office to show your support for survivors of pedophilic abuse. It's quite another to condemn any consideration of an unpopular possibility by voting through an act of Congress that pedophilia *must be* as harmful as the public generally believes, no matter what the studies showed about the reality.

What seemed to bother Sher most, though, wasn't the stupidity of Congress. It was the way the American Psychological Association had handled the matter. In spite of being reasonably worried that the controversy might be used to cut federal funding to such institutions as the National Institute of Mental Health, initially the APA had done a good job keeping the politicians at the gate. Despite calls for the editors' heads, they were not removed by the APA, and the APA kept Sher and Eisenberg apprised of what was going on. One of the senior staff even made a point of calling to acknowledge the stress Sher and Eisenberg had to be experiencing.

But then, as DeLay increasingly threatened to go after the APA itself, Raymond Fowler, the association's chief executive officer, caved. He wrote what came to be known as the capitulation letter, assuring Delay that "the article included opinions of the authors that are inconsistent with APA's stated and deeply held positions on child welfare and protection issues." Meaning what, exactly? That the APA's "stated and deeply held positions on child welfare" included that all victims of pedophilia must be profoundly and immutably harmed?

More problematically, Fowler announced that the APA would

arrange for an independent review of the Rind paper, an unprecedented move and a seeming admission that if a paper's PR was bad enough, the normal scientific review process could be subverted in the service of politics. Sher recalled to me that the APA turned to the American Association for the Advancement of Science (AAAS) to do the review, only to have the AAAS do what the APA should have done: defend the scientific process from political meddling. Irving Lerch, chair of the AAAS Committee on Scientific Freedom and Responsibility, dressed the APA down:

> We see no reason to second-guess the process of peer review used by the APA journal in its decision to publish the article in question. . . . We believe that disputes over methods in science are best resolved, not through the intervention of AAAS or any other "independent" organization, but rather through the process of intellectual discourse among scientists in a professional field.

Lerch also suggested in his letter that the APA might have done more to correct the public mischaracterizations of the Rind paper, rather than implicitly repeating them through capitulation.

Many saw Fowler's letter on behalf of the APA as selling out not only the Rind paper's authors, editors, and reviewers, but science itself. You have to wonder if Fowler or his staff was ashamed of what he did. Up until this point, the APA had made sure to keep Sher and Eisenberg apprised of what was going on over the Rind paper. But news of Fowler's capitulation came not directly from the APA, but through an improbably circuitous route: Sher learned of it from Eisenberg, who learned of it from her editorial assistant, who learned of it by hearing Dr. Laura on the radio trumpeting her little victory over the APA.

This was hard to swallow. So I swallowed a bit more of my drink and remarked to Sher how odd I thought it was that people would be so angry to hear that not every victim of pedophilia had had his or her life utterly ruined. It seemed to me the Rind paper contained a bit of good news for survivors, namely that psychological devastation need not always be a lifelong sequela to having been sexually used as a child by an adult in search of his own gratification. But simpler stories of good and evil sell better.

Remembering the whole fiasco, Sher recalled to me how the process had been rigged. The Congressional resolution condemned together both pedophilia and the Rind paper, so as Sher and Eisenberg later noted in a written reflection on the whole mess, "One could not vote in favor of the [Rind] article without voting *for* pedophilia." If you wanted to try to distinguish pedophilia and the scientific process, abstaining from voting was the best you could do. Surely DeLay purposely set it up that way.

I didn't bother asking Sher if he'd be voting for Obama.

CRAIG PALMER'S OFFICE had the oddest homemade doorbell I'd ever seen, one that reminded me of the *Winnie the Pooh* story in which Owl accidentally walked off with the donkey Eeyore's tail and turned it into a bell pull. When Palmer opened his door, I realized what was up. His office consisted of two rooms, an anteroom and an interior office, so that if he happened to be working in the inner room at his desk, he might not hear a person knock at the outer door—hence an elaborate contraption that allowed a visitor to pull on a long string that would ring a little bell hanging within earshot of Palmer's desk.

Mike Bailey's son, Drew, had been particularly keen on my talking to Craig Palmer. Before my trip, Drew had checked several times to make sure I knew about Joan Roughgarden's review of the book Craig had coauthored with Randy Thornhill, *A Natural History of Rape*, a book that explored biological explanations for forced sex. Roughgarden was the trans woman scientist at Stanford who had become one of my most vocal critics following my work on the Bailey controversy. Months before I had gone to Columbia, Drew had sent me the "biology of rape" review Roughgarden had published in *Ethology*, and I assured Drew I remembered it. You don't easily forget an essay in a scientific journal that calls for authors of a scientific monograph to swing in the wind. (Quoth Roughgarden: "Thornhill and Palmer are guilty of all allegations and they deserve to hang. But before stringing them up, let's reflect.") But even Roughgarden's contempt for these guys would not have made me like them if they had actually said what they'd been accused of saying: that rapists should be excused and forgiven because their genes made them do it and that raped women had been asking for it. Of course, they hadn't said that.

What Randy Thornhill and Craig Palmer *had* said was that rape has a sexual component to it—that contrary to the claims of some feminists, rape isn't merely an expression of unadulterated power. Thornill and Palmer marshaled evidence suggesting that some kinds of sexual coercion in some species, including humans, may increase the likelihood of reproductive success of some males. They also collected evidence showing that human rapists in general tend to be interested in women of childbearing age whom they find sexually attractive. Notably, Thornhill and Palmer took very seriously the harm caused to women by rapists and argued that truly caring for victims of rape meant taking seriously possible biological contributions to sexual coercion. While their work might help to explain rape—and, they hoped, even prevent and prosecute it—they certainly did not excuse, condone, or forgive rape. Contrary to Roughgarden's assertions, they did not provide "the latest 'evolution-made-me-do-it' excuse for criminal behavior."

Craig had told me in advance of our meeting that he didn't much enjoy thinking back to what happened when the book had come out but that he had kept a mess of papers related to the controversy in a filing cabinet. He said that, given Drew's recommendation of me, I was welcome to go through the collection with him. As we settled in to talk in his office, he told me the story of having been in a class a year or two before, talking about the controversy over his work, and finding that this one grad student named Drew Bailey was thoroughly engaged. Finally he realized that the kid was Michael Bailey's son—*that* Bailey.

I laughed and asked Craig if he was aware that, if you Googled "Thornhill and Palmer," one of the first hits you got was a page from Lynn Conway's Web site attacking Thornhill and Palmer and trying to tie them to the Bailey controversy. Craig apologized that he didn't know what I was talking about. I explained as best I could.

Although Craig's office counted as a model of cleanliness and order for an academic suite, the file drawer in which he had collected writings on his own controversy was a total mess. The papers were stacked horizontally in some places, shoved in vertically in other places, some in folders or envelopes but many without. It was obvious that this portion of Craig's life had been chaotic and that he had literally just put it all away and moved on. Now, as he went through the jumble, trying to create logical stacks for me on his table, he tried to explain what had happened.

Like Mike Bailey, Craig Palmer and Randy Thornhill weren't naive, and they weren't shy. They knew that when they set out to collaborate on their mutual interest and to publish together on the biological bases of sexual coercion that the work would draw attention and also some ire, but they had no idea what they were really in for. The first inkling that something was up came when the two of them went to Boston for a meeting at MIT Press, the outfit publishing their book. Craig and Randy thought that they were going to discuss how the book would be promoted, but when they got there, they were suddenly informed that they had to present a lecture on the work to a group of people the press had assembled. Craig recalled to me, "We walk in, and there were 50 or 60 people in this room." The authors were understandably bewildered. They'd never heard of such a thing (nor had I). Randy let Craig take the lead, and as Craig recalled, he jotted down a bunch of notes on a legal pad and went from there. Though there was plenty of hostility to the project, Craig and Randy felt the sum-up went pretty well. They were able to handle all the questions, however misdirected, but it was a disturbing situation nonetheless. Craig told me, "It was clear that news of the book had spread around MIT, and the people there were basically protesting the publication of the book."

Then in January 2000, Randy and Craig published a summary of their forthcoming book in an article entitled "Why Men Rape" in the *Sciences*, a magazine of the New York Academy of Sciences. All hell started to break loose, and things only got worse when the book came out in April. Craig told me,

> The very first media attention I knew of [came via] a phone message from a friend from high school, saying that Rush Limbaugh was talking about my book on the radio. Obviously not the typical phone message. First I thought, how did Rush Limbaugh hear about our book? Then I wondered what in the world Rush Limbaugh thought of it. Then I wondered what happened to my friend that he was listening to Rush Limbaugh.

Craig went on:

> It was interesting, because as I thought about it, I thought I could see Rush Limbaugh going either way on it: he could like it because we were challenging the feminist explanation of rape, or he could dislike it because we take an evolutionary approach. But it turned out he was criticizing it because he thought we were trying to excuse Bill Clinton's behavior.

Long after Limbaugh lost interest, the heat kept up. One criticism after another came flying at Thornhill and Palmer, in the popular media, in the presses of the intelligentsia, and in the mail. As Craig showed me, the majority of these criticisms attributed to Craig and Randy various ignorant and obnoxious claims that they had never made. For example, in *Time* magazine, Barbara Ehrenreich suggested that Thornhill and Palmer seriously downplayed the amount of harm done to rape victims, even though the book takes that harm very seriously, even attempting to quantify it and make sense of variations in levels of harm. (Perhaps like Dr. Laura in the Rind case, Ehrenreich just couldn't wrap her head around anything other than the classic story of sexual assault in which the victim is always irrevocably devastated.) A letter writer to the *Los Angeles Times* assumed that because Thornhill and Palmer said rape was sexual, they were also labeling it normal. The *Nashville Tennessean*'s headline called the work a "'Can't Help It' Theory," while the *Manchester Guardian* similarly announced "The Men Can't Help It," as if Thornhill and Palmer had concluded that men amounted to pathetic slaves to their evolutionary histories. Meanwhile, the *Toronto Globe and Mail* ran angry letters under the title "Are Men Natural-Born Rapists?" as if that was exactly what *A Natural History of Rape* had concluded, the reality of the book be damned.

The feminist writer-activist Susan Brownmiller seemed particularly furious, and no wonder. In their work, Randy and Craig directly took issue with Brownmiller's highly influential opinion that rape is essentially about power and domination, not lust. Thornhill and Palmer acknowledged that the treatment of raped women in courts and in society had greatly improved since the time of Brownmiller's bold work, and indeed it had. Brownmiller and other feminists had radically changed the public story of rape by reframing it as symptomatic of a pandemic disease—patriarchal misogyny. By talking about how rape is used as a tool of power and intimidation, by steadfastly seeing rape as part of cultural systems that oppress women, Brownmiller and others had changed many harmful and entrenched cultural assumptions about rape. No longer could someone easily get away with blaming a rape victim for what she was wearing; no longer was *she* the one to be on trial.

Thornhill and Palmer shared Brownmiller's desires for an end to rape and for compassion and justice for rape victims, but they argued that Brownmiller's account of rape as primarily being about power didn't match the facts. Men seeking power over women could find it in a number of ways, but the *choice* to rape a woman and especially the ability to sustain an erection during a rape suggested, at the very least, significant sexual arousal. Denial of that reality, Randy and Craig argued, would only lead to more harm to women.

As Craig recalled the public battles with me, he pulled out an example of Brownmiller's influence. He handed me a pamphlet distributed by the University of California–Davis's Rape Prevention Education Program, a branch of the university's police department. Here's some of what it said:

> FACT: Sexual assault is an act of physical and emotional violence, not of sexual gratification. Rapists assault to dominate, humiliate, control, degrade, terrify, and violate. Studies show that power and anger are the primary motivating factors. . . .
>
> FACT: Sexual assault victims range in age from infants to the elderly. Appearance and attractiveness are not relevant. A rapist assaults someone who is accessible and vulnerable.

Craig looked visibly disturbed as he read this stuff. Was it really ethical to suggest to a woman that she didn't need to be concerned about how attractive a potential rapist might find her when she was in an environment where she was vulnerable? And did it really make sense to suggest that rapists were never motivated by sex?

Craig explained to me that these were just the sorts of claims that had led him into this work. In the mid-1980s, Craig had been studying anthropology in graduate school at Arizona State University, but he'd decided to give up on it. "I dropped out of graduate school because postmodernism was then coming into academia," Craig explained to me, "and it didn't seem worthwhile to go through all of the hard work of science and present all the evidence, just to have it dismissed by [someone] saying, 'Well, that's just your narrative.'" So he dropped out of school, moved up to Maine, married, bought a house, and settled in as a lobsterman.

But circumstances around a rape combined to bring Craig back to academia. The year before Craig moved to Maine, in the Arizona neighborhood where he lived, a neighbor's daughter had been kidnapped and murdered. "They caught the guy they thought had done it," Craig recalled, "and said they had all kinds of forensic evidence that he had committed the crime. About three days after the body was found, I remember there were two headlines on the same page. One headline . . . said something like 'Autopsy Determines Victim Was Sexually Assaulted.' Which I think everyone who heard about the case expected. But on the same page, the main headline said something like 'Still No Motive Found.'"

Craig recognized that this juxtaposition reflected the standard dogma about rape—that rape (and thus also kidnapping and murder done to facilitate a rapist's aims) were not explainable simply as a sexual act gone evil. Before he gave up on school, Craig remembered telling his mentor in Arizona, "If I come back, the claim that rape isn't sexually motivated might be one obviously wrong social science explanation worth challenging."

While living in Maine, Craig got a call from an assistant district attorney in Arizona. About this, Craig told me,

> My first thought was that I must have forgotten to turn in some library book that was now a year overdue. But the assistant DA said the rape/murder [of the neighbor] was coming to trial, and his job was to contact everyone who lived in that part of town to see if he could find anyone who had seen an angry interaction between the girl who was raped and murdered and the man who was accused.

Craig had nothing to offer. "I was curious about why he would need someone to report that specific thing, so I asked him, . . . 'Couldn't you just argue that the guy was sexually attracted to this girl and he knew she would never have sex with him willingly?'" Couldn't the DA reasonably postulate that the motive was sexual and the murder had been committed to cover up the crime?

The DA answered that that was basically what they had tried to argue, but "the defense said something like 'scientists had proved rape is not sexually motivated. Instead it's motivated by a desire for violence, control, or power.' So that's the motive that had to be established." Craig found himself thoroughly frustrated. Did it really make sense to talk about rape as if it were a nonsexual act—especially when such a poorly evidenced claim gets in the way of bringing rapist-murderers to justice? Clearly, the populist dogma "could even let a murderer and rapist go free," he said to me. "So I asked my wife if she'd mind if I went back and finished my PhD in order to write a dissertation challenging that explanation of rape." His wife did not mind, and Arizona was willing to take him back and backdate a leave of absence. Craig finished his last semester of course work and then returned to Maine to write his dissertation on rape while paying the bills in lobsters.

In writing about the biological bases of sexual coercion, Craig inevitably encountered the work of the zoologist Randy Thornhill, who was interested in the same topic. In his studies of scorpionflies, Randy had found plenty of evidence that male scorpionflies prefer consensual sex, i.e., sex in which the female participates, typically in response to male presentation of a nuptial gift, something along the lines of a dead insect or a mass of hardened saliva. But if a male is unable to get a female to cooperate sexually—if, for example, he can't get his legs on a gift—he will resort to forced sex, using a grabbing organ that appears to have evolved for just this purpose. Craig and Randy understood that humans differ radically in many ways—no one's getting this girl into bed with a gift of petrified spit, and men don't have a specialized rape-facilitating organ—but Craig and Randy also understood the value of recognizing that human sexuality has *evolved*. Some men's rape of women might therefore be explainable with the tools of evolutionary biology.

In their early (and later) discussions, Craig and Randy sometimes disagreed about evolutionary explanations for coerced sex. Craig thought sexual coercion was likely to generally represent a by-product of evolution—an accidental side effect of evolutionarily successful adaptations—whereas Randy was inclined to see evidence for adaptionist explanations. But the two realized they agreed more than they disagreed, and so they decided to write *A Natural History of Rape* together, a book that would work through massive amounts of data on sexually coercive acts in humans and other species.

As one might predict, when the media frenzy set in around the book, Craig and Randy had the typical experience of those who challenge conventional wisdom. Commentators took the existing stories of good and evil, good guys and bad guys, acceptable claims and unacceptable, and tried to fit Thornhill and Palmer into those preexisting slots. Since Thornhill and Palmer seemed to be saying unacceptable things about rape—it *does* matter whether a victim looks sexually attractive to a rapist; rape *is* often about sex; biology *does* contribute to sexual coercion—that meant they had to be the bad guys. Thus, in the crunching of the daily media machinery, they were magically transformed into misogynistic apologists for rape. Then all you had to do was put into their mouths the words Bad Guys say about rape: "The woman was asking for it, and the guy couldn't help himself." And so came the hate mail and the threatening phone calls. About those, Craig told me, "Let's just say I learned the legal line that separates official death threats from run-of-the-mill nasty e-mails and letters."

The messages left on Randy's answering machine were so frightening that a local sheriff did him the favor of recording the outgoing message for him in a very macho voice. His message indicated that the speaker was a law enforcement official, said that the call was being recorded, and reminded people that it's against the law to call people up to tell them you're going to kill them. The business about the police recording the call wasn't true, but it helped. Randy (and his kids) no longer had to hear what people wanted to do to this Thornhill guy. Meanwhile, on Craig's end, "Things were so bad that the police told me to take some precautions, like checking my car for car bombs every morning and varying my routine. I was even provided a special parking place on campus they thought would be safer."

Um, well, OK. I'd definitely found someone who'd had it worse than I. Humbled, I asked Craig what sustained him during this time. It obviously helped that he had to take care of his family; when that focus is required of a family oriented guy like Craig, there will be a certain lifesaving structure to one's days, even in the face of insanity. (I knew that firsthand.) I suggested to Craig that it probably also helped that he and Randy were in this pickle together. Craig agreed. Not only did being "Thornhill and Palmer" mean built-in peer support, it also meant that Randy and Craig felt a responsibility to each other to see the business through—to stay professional, rational, and unafraid. But Craig mentioned one more thing that kept him going: the sense that he was right in a way that ultimately would help women. He had more and more reason to believe that was true because he was getting messages from rape survivors who told him they appreciated his understanding that the men who assaulted them had done so for sexual gratification. Some were even thanking Thornhill and Palmer in public.

In the Lifestyles section of the *Dallas Morning News*, Elizabeth Eckstein began her op-ed this way:

> Finally. Finally, somebody is coming around to my way of thinking on the motivations of rape. I can say this because I survived an aggravated sexual assault by a serial rapist and, more important, two years of post-traumatic stress syndrome that included an exhausting state of hypervigilance, sudden panic attacks, yelling at God and the cold clench of fear in my gut. I also was consumed with an obsessive (some would say

> unhealthy) need to know why. Why me? Why him? Why rape? So I tried to find out. During my quest, I came across a lot of people who liked to quote the so-called experts and say things such as, "It's a crime of violence, not sex" and "It's a control thing." Boy, did I hate those people. In my mind, they were wrong. I used to reply to those sorts in a real catty fashion. "He didn't force me into the kitchen to break all the dishes. He didn't make me smash all the furniture in the house. He made me have sex with him against my will. Sex, people, sex at gunpoint. Choice absolutely and totally removed from the equation. An act, typically one of love, reduced to its lowest and ugliest form."

Craig made sure I saw this in his stack of photocopies. I read it quietly and remarked to him how bizarre it was that we had reached a point at which we have to argue that an act that involves an erection and typically results in an orgasm is a *sexual* act.

"Exactly," he answered.

Eckstein was not alone in thanking Thornhill and Palmer for challenging the Brownmiller construction. In an interview with the *Boston Herald*, Jennifer Beeman, director of the Campus Violence Protection Program at UC–Davis—yes, the very university whose "rape is not about sex" pamphlet I quoted above—"said she hopes the article and book [by Thornhill and Palmer] will force scientists, social scientists, women's organizations and rape experts to do some soul searching. 'For so long our mantra has been "It's about power, not sex,"' she said, 'that I think we're afraid to admit it might be about both.'"

At some point, Craig pulled an envelope out of the stack, and from the envelope pulled out a four-page handwritten letter from a guy serving time in a federal penitentiary in the South. Craig skimmed it, paused, and handed it to me, saying I needed to be sure that if I used this letter, I not identify the writer. The handwriting was neat, the prose clear, and the writer pretty well educated. He had read or heard of

Randy and Craig's article in the *Sciences*, and he was writing to give Randy his own opinion:

> I know from repeated first hand experience that sex really is the central motivation to rape. Although this may not always be true with all offenders, or even all cases of adult rape, I know from my own self introspection through offender programming treatment, and from other adult offenders I've been in such programming with, that sexual attraction and instinctual sex urges acting as biological imperatives strongly motivated acts of rape (strangers/adult female). It's frequently confessed.

The convict went on to explain about how rapists—presumably just like him—pick off females as attractive and "available" targets. He agreed with Palmer and Thornhill on this: "A *dumb* myth is that rapists go after *any* female." The writer went on: "Although from a therapeutic view it is of course important that an offender in no way get to abrogate his guilt by placing blame on the victim's real or imagined provocative behaviors, the school of thought [that says rape is not about sex] stymies and downplays the existence of these powerful motivations."

How, the letter implicitly asked throughout, can rape be successfully prevented, and rapists treated or at least adequately controlled, if we deny the reality of men like this? The correspondent, a man hopefully locked up behind bars for a very long time, was clearly moved to commit a heinous crime by a pathological kind of lust. But a *lust* nonetheless.

Reading this letter, I found myself going still with fear. Or rather, a combination of fears: One part a fear of rapists like this man, who will see a sexual solicitation in a woman's bending down to pick up her keys. One part a fear of ideologies like feminism—ideologies that might lead reasonable, progressive people like me to accidentally hurt someone for the sake of a more palatable or more useful argument. And one part a fear that anything I say can and will be used against me, as had happened to Craig.

TALKING TO CRAIG in his office about his frustrating encounters with my fellow feminists soon had me involuntarily mulling over what had happened to me at the annual meeting of the National Women's Studies Association (NWSA) almost five months earlier. It was there that, for the first time in my life, I found myself wondering whether I should (or could) really call myself a feminist.

When the call for conference proposals had gone out for the 2008 NWSA conference, to be held in Cincinnati, several colleagues had written to me upset about a post on a major Women's Studies e-mail discussion list from a trans woman graduate student named Joelle Ruby Ryan. Ryan was putting together a panel for the meeting and was seeking participants for what would no doubt be a scathing criticism of me and Bailey. The title of the planned session? "The Bailey Brouhaha: Community Members Speak Out on Resisting Transphobia and Sexism in Academia and Beyond." This was obviously payback time for my exposé of what Andrea James, Lynn Conway, and Deirdre McCloskey had done to try to shut Bailey up. Not too surprisingly, Ryan's work was funded by an LGBT foundation grant for which Conway functioned as Ryan's "mentor."

The call for proposals lumped me and Bailey together, suggested that my work contributed to a "chilly climate" for transgender academics, and so forth. No wonder colleagues on the list were writing to me to ask me to defend myself. So I hopped on this e-mail list and attempted to point people toward my peer-reviewed article and the *New York Times* coverage of it. I also pointed out that anyone could submit a response and have it published right alongside my article in *Archives of Sexual Behavior.* I threw in a note that the call for panel participants was riddled with problematic claims.

Knowing I'd basically been set up, I also wrote to the programming committee of NWSA and asked that I be given time to respond to the panel. No deal. I wrote a proposal to contribute to the panel, sent it to Ryan, and was told there was no room for me. I saw in the coming plans for the session that there *was* room for Andrea James and two other trans women—with Lynn Conway functioning as the "session advisor." Finally, I submitted to the NWSA conference organizers my own proposal for a paper, a comparison of techniques used in the intersex rights movement and in the Bailey book controversy. The conference organizers granted me a slot in a random session earlier in the day.

In Cincinnati, in my allotted fifteen minutes, I pleaded with audience members to attend the later session, the session dedicated to taking me down, and yet to not simply believe what they would hear. Don't believe what you have not seen evidence for, I told them. I asked people to think about the importance of evidence to issues of identity rights. I talked a little bit about why scholarship is not the same as activism. The audience looked incredibly uncomfortable. In a first for my public talks, no one had any questions for me afterward.

A few hours later, when I walked up to the room where the session Ryan had organized would be, I found a legalistic note on the door, saying that anyone entering automatically gave Andrea James the right to videotape them and use the recording for her own purposes. I backed up, went over to the conference organizers' stand, and asked them to come deal with that. One of the organizers came and took the note down. I then walked in, and saw not only Andrea James, but indeed Lynn Conway herself, standing behind the video camera ready to tape. And who was at her side but Juanita, she whom Conway had used to try to ruin Bailey on the sex charge.

Andrea James began the session by explaining that the note on the door really just meant that they would videotape the panelists for their own purposes. Uh huh. I decided I would say nothing during the session, afraid that anything I said would be clipped and twisted in an edited recording. The whole scene was maddening. Panelists repeatedly defended James's online abuse of Bailey's children, never explaining to the audience what James had actually suggested in conjunction with those photos—that Bailey might have sodomized his own children. They instead focused on the most outrageous lines of Bailey's book, without any context—perhaps lest they let any novice grasp the real issue, autogynephilia? They never mentioned that trans women who had agreed with Bailey had been silenced by harassment and threats—including by one of the panelists now being given a spot here at a legitimate academic conference. Of course they didn't delve into the falsity of the charges against Bailey or my refutation of them. What could they say? My findings were all documented.

Then of course there were predictable claims about *my* position: A non-queer person could never understand the reality of queer people. My work could silence trans women in the academy, women who allegedly lacked the privilege I allegedly had. One identity card after another was thrown down—which only made sense in a "feminist" room where you win simply by having the most identity cards. I found myself thinking that Women's Studies is about as sophisticated a game as Go Fish.

Most maddening, one of the panelists actually had a few interesting critiques of my work. For example, she took me to task for not adequately exploring the ways in which Bailey deployed the socially powerful term *science* even while putting forth sometimes oversimplified accounts of identity. She complained that, in my write-up of the Bailey history, I did not accord transgender people the kind of humanizing narrative attention I had accorded intersex people in my earlier work. I found myself going crazy with frustration that I could not, in this audience, engage or even acknowledge interesting criticism, because of the taping by James and Conway, because of all the stupid politicking being allowed to happen in the name of feminism. And what kind of feminism?

During all this, one young woman seated next to me remarked to her friend that this Dreger woman sure is terrible. I leaned over and whispered to her that *I am* that Dreger woman and that I did not recognize the person the panelists were describing. She turned away as though she had just met an armed skinhead wanted for murder. I just sighed.

At the end, there was a little time left for Q&A. I turned around to see the first person called upon: a tall trans woman, sitting near the back of the room. *Here we go,* I thought, *more piling on.* I braced myself for the next blow. Instead, the woman stood up and said this:

> [I am] Rosa Lee Klaneski, [from] Trinity College. I cite Alice Dreger's academically rigorous work all the time in my own work. She doesn't know who I am, but I know who she is. And I am just wondering—and I'm a transgender person myself—what gives any transgender person the right to abrogate someone else's First Amendment right to freedom of speech just because they hold an unpopular minority view? In my opinion [regarding] the person that you are arguing against [meaning Bailey], I completely agree with you. Bunk. Ridiculous science. And should be classified as such. I got that. What gives us the right to censor it just because we don't like it?

Stunned, I turned back to see how the panel would respond. Predictably, they argued that the panel didn't constitute censorship. How was this panel censoring people like Bailey or me? But I thought, *come on.* The note on the door, the Web pages, the video camera, and what so many sex researchers had said to me: that no one in sex research will touch male-to-female transsexualism with a ten-foot pole anymore. Which must have been just what Conway meant to do. And there was Conway, "mentoring" Ryan and taking it all in.

Then suddenly the session was out of time, and it seemed pretty much over. I went to the back of the room to Rosa Lee Klaneski, and shook her hand.

"You're right," I said, holding back tears, "I don't know who you are, but I would like to know you. Can I buy you a drink in the bar downstairs?" She answered with a smile that I could certainly buy her a cranberry juice with soda water and lime. At that moment, out of the corner of my eye, I saw Andrea James go up to my friend April Herndon, my former graduate student who had also worked on ISNA's staff during my last year. James had clearly been doing her research; she seemed to know just who April was and to know of her relationships to me. I'd been so careful not to sit with April, lest she get in their sights, but my decoy had not worked. James seemed to be trying to corner her into saying something politically incorrect. The camera must have surely been pointed our way. I pulled on April's sleeve and told her to just walk away. Otherwise we'd all be on YouTube by evening, positioned as oppressors to all trans women.

Rosa, April, and I left the conference room and started to walk down toward the hotel bar with a fourth woman whom we knew from intersex work and who had insisted on coming with us. But as we made our way, James suddenly came up to me.

"Alice, honey," she said to me, towering over me, "I'm not done with you. In fact, I haven't even started with you." She said she was still going to prove Bailey had lied in his book. "I'm going to ruin your career."

In a split second, Rosa stepped between us, and calmly spoke as if to me, though clearly actually speaking to James. "Alice," she said, "the legal definition of assault does not require that a person touch you. You can call the police right now and report assault."

At that, James hastily stepped away.

We went down to the hotel bar. It was the middle of the day, and yet after arranging nonalcoholic drinks for the other three, I ordered a gin and tonic for myself, and then another. As the sedative washed over my brain, Rosa told us about herself, mentioning that she had a degree in women's studies but was tired of the bullshit of the field. She was finishing her master's degree in public policy now, writing about her own experience of trying to change the sex on her driver's license without being forced into medical procedures she didn't want. She had some radical ideas about how to harness capitalism to push for transgender rights. Enough of the liberal feminist and queer rights rhetoric; it was time to use the existing economic system and work through for-profit institutions to make the world safe for trans people. In the meantime, while she finished her master's, Rosa was working in the pawn industry in Connecticut. I got the sense that the guys in the business had had to accept her as a transgender woman because they had enough business smarts to know they needed her.

Rosa told me that she was also an up-and-coming poker player and was working on a nutritional supplement designed to help players concentrate better. She slipped me a sample packet of her product across the table, and I put it in my bag, wondering if I should pop it now. I wanted to be able to focus—to remember this. For there Rosa sat, rattling on, just so funny and calm and kind and independent-minded and smart and brave.

When it came time to leave, Rosa said a warm good-bye and added, "Seriously, you let me know if you need anything." With her positive reputation in the pawn business, she had, you know, *good connections* with guys who know how to handle little problems. "You just let me know if you need anything. A sympathetic ear. A little *protection*." She paused. "A slightly used big-screen TV?" Her mischievous smile made me wonder if she was serious.

AT THE UNIVERSITY OF MISSOURI, after talking for hours with Craig Palmer, I went on to talk to the other interviewees I'd arranged to meet, but I found it hard to concentrate. I kept doing a really weird math: What's worse, having your work denounced by an act of Congress or trying to help prevent rape only to be accused of fomenting it? What's more terrifying, being charged with having sex with a research subject or getting a lesson in how to check for bombs wired to your car? And who is the real feminist, the one who reflexively sides with people who've been historically downtrodden or the one who does so only after checking the facts?

And then a reactionary calculus question emerged: Is there anything too dangerous to study? Should there be any limits? What if, in order to prove how important truth seeking is, we made a point of studying the most dangerous ideas imaginable? What if we even really studied *race and IQ*?

Yeah, apparently I was now getting drunk on the idea of absolute intellectual freedom. I mean, I could see that no good and much harm could come out of certain scientific pursuits. (Oh, like studying race and IQ.) And yet, I kept thinking: What if we became unafraid of all questions? Unbridled in our support of the investigation of "dangerous" ideas? What if we came together in the ivory towers, barricaded the doors, and *looked at the skies*?

Never before that trip to Columbia had I felt a burning sense of being an academic. Never before had the profession felt to me holy in the way it was beginning to feel now. I found myself becoming bizarrely sentimental about donning my PhD robe and hood, those leftover symbols of the monasteries from which universities had emerged. Those monks had been about a supernatural truth. We must be all about earthly truth. And our pursuit of the truth *would be* our pursuit of justice, our defense of democracy. We would not allow the DeLays of the world to stop us. We would not put up with the American Psychological Association and the National Women's Studies Association kowtowing to identity politics. The identity that mattered to us would be our identity as academics, as truth seekers.

And of course we would not naively believe that any of us could find the truth alone. We must honestly assess each other's work. This, I had long taught my students in history-of-science classes, was the genius of science: the ideal of peer review. The light of many minds. Not coincidentally, this was also the genius of modern democracy, I suddenly realized in Missouri—the Show Me State. The Enlightenment brought us both science and democracy. The Founding Fathers had understood the usefulness of the scientific review model. The three-branched system of government, with its checks and balances, the jury system, a Supreme Court with multiple justices approved by multiple representatives—these institutions were meant to do just what the review process of a good journal is meant to do: Weed out the bad, leaving the good.

But could we do it? Could we manage it in an era of moneyed interests—defense contractors and drug companies and oil monopolies (and Conways) financially manipulating the systems in ways we couldn't even see? Could we do it in an era when the Internet allowed people like James to create "truth" through clever marketing strategies? Could we take back an academy that had allowed itself to become so beholden to external funding, identity politics, twenty-four-hour news cycles, and conniving legislators?

As we had talked in his office, Craig Palmer had said of his experiences: "From all this, I've learned more about the human species and how it can do things like lynch mobs and genocide and stuff. I'm not sure I'm glad to have that knowledge. One of my colleagues asked if the experience had lowered my view of the media. And I said no, it's lowered my view of the species."

Yet as I sat in the Columbia airport in the predawn dark of Halloween 2008, waiting for my early morning flight home, I felt strangely hopeful. I meditated on the actions of one woman long ago, a woman whose name I didn't even know: the woman who had been Craig Palmer's dean in Colorado when all hell broke loose. Craig had been in a nontenure line, utterly vulnerable. With all that bad publicity, all the trouble with the threats of violence, it would have been easy for his university administrators to cut their losses by cutting Craig loose. It was not as if they weren't getting letters calling for him to resign or be fired. And what had his dean done? *She had defended his academic freedom.*

They called for the passengers on our flight. I went through the metal detector and out the door that led to the tarmac where our plane waited. In the earliest light of the day, I looked up to see a great big plane right next to our little one, and I stopped in stunned surprise. It was Obama's campaign plane, *Change We Can Believe In.*

"Can I take a picture for my son?" I asked the Secret Service agent standing there.

"If you hurry," he said, taking my suitcase and smiling broadly.

In four more days, I thought to myself, *the people will peer-review.* And this man will be our president—this intelligent, well-read man, this man who speaks of restoring science to its rightful place. Restore the scientific process; restore democracy. This is what we needed—to develop a core identity as *American academics*, the people who would make sure a Galileo was never again put under house arrest for making challenging claims about who we really are. Make people understand the difference between a self-serving personal narrative and an empirical study that had undergone rigorous peer review. Teach people why they all should want to be like academics saying "show me" at every step of the way.

I stuck my camera phone in my pocket and took back my bag from the Secret Service agent. And I stepped onto that little prop plane positively high.

CHAPTER 5

THE ROT FROM WITHIN

THE STORY I HAD BEEN TOLD about Mike Bailey and Craig Palmer and so many other white straight male scientists accused of producing bad and dangerous findings, the story I had willingly heard as an academic feminist in the humanities, was that these guys were just soldiers of the oppressive establishment against which we good guys had come to fight. *They* came from old dogma about human nature; *we* came from progress and social justice, and so we had to win. But here I was faced with the fact that not only were these scientists politically progressive when it came to things like the rights of transgender people and rape victims, they were also willing to look for facts that might get them in hot water. They very much cared about progress in social justice, but they cared first about knowing what was true.

That didn't mean that these scientists (or I, or anyone else) existed without bias. It didn't mean their work wasn't shaped and sometimes tainted by politics, ideologies, and loyalties. But it did mean they tried to adhere to an intellectual agenda that wasn't first and only political. They believed that good science couldn't be done by just Ouija-boarding your answers. Good scholarship had to put the search for truth first and the quest for social justice second.

In Missouri, I realized that there's a practical reason for this order: Sustainable justice couldn't be achieved if we didn't know what's true about the world. (You can't effectively prosecute and prevent rape if you don't understand why, where, and how rape happens.) But there was also a more essential reason for putting the quest for truth first: *it was who we scholars were supposed to be.* As the little prop plane flew from Columbia, Missouri, toward the sunrise, of this I was sure: We scholars had to put the search for evidence before everything else, even when the evidence pointed to facts we did not want to see. The world needed that of us, to maintain—by our example, by our very existence—a world that would keep learning and questioning, that would remain free in thought, inquiry, and word.

Nevertheless I knew many of my colleagues in the humanities would disagree. I could practically hear them arguing against me, as if they were seated all around me in those cramped fake-leather seats, yelling to be heard above the churning propellers. *We have to use our privilege to advance the rights of the marginalized. We can't let people like Bailey and Palmer say what is true about the world. We have to give voice and power to the oppressed and let* them *say what is true. Science is as biased as all human endeavors, and so we have to empower the disempowered, and speak always with them.*

Involuntarily shaking my head, I argued back: "Justice cannot be determined merely by social position. Justice cannot be advanced by letting 'truth' be determined by political goals. Only people like us, with insane amounts of privilege, could ever think it was a good idea to decide what is right before we even know what is true. Only insanely privileged people like us, who never fear the knock of a corrupt police, could think guilt or innocence should be determined by identity rather than by facts. Science—the quest for evidence—is not 'just another way of knowing.' It's a methodical process of checking each other, checking theory against experiment, checking claim against fact, and fact against fact. It isn't perfect, but look what it has gotten us: antibiotics, an explanation and a treatment for AIDS, reliable histories of the Holocaust, DNA-based exonerations of those falsely accused of crimes, spaceships on the surface of Mars—hell, the plane we're flying in now."

Where would we be, I wondered, if the pope had ultimately won out over Galileo, if he had succeeded in using his self-serving Catholic identity politics to forever quash Galileo's evidence that the ancients and the Bible were wrong about the Earth? Power plays as morality plays, whether by popes or feminists, are just that—*plays.* I longed for the real world, longed to get away from discussions about "representations" of reality. I longed to pick apart each history to know what's true, to have my work judged by others, to find evidence that an idea is right *or wrong.*

As we flew on, political loyalties that had once felt grounding to me now felt like heavy weights I longed to drop away. There seemed a promise, with this lifting, of being able to see much further, much clearer. Surely Galileo must have felt something like this combination of longing, doubt, and hope when he looked up and realized fully that there were no crystalline spheres of the kind the ancients had said affixed and turned the planets and stars around the Earth. No longer any rigid spheres in his world, just infinite soft space, infinite potential for discovery, for wonder, and for trouble. We are not standing on a still and special earth created by God in the very middle of a perfectly round universe, but on a lumpy twirling planet, a planet bright from reflecting sunshine just like the other planets, circling around a pimpled sun, spinning along in the middle of a comet-littered nowhere, somewhere in a vast and messy universe—a world that might now become known as never before.

No wonder Galileo learned what he needed to learn in order to construct good telescopes. No wonder he spent so many of his days struggling to make those visual portals better and better. Religiously speaking, the pope had the power to stop Galileo from achieving salvation; he could excommunicate him, mark him as a bad soul forevermore. But I suspect that, in his heart and through his telescopes, Galileo had already achieved the kind of salvation that matters to the seeker. He had achieved a philosophy that truly liberated him, and then also us, his enlightened descendants. The pope might claim to gatekeep for God, but in truth, even the pope couldn't stop Galileo from climbing into the heavens to pull down facts and bring them back to Earth. And Galileo knew that he had achieved not only a physical truth, but a metaphysical liberation of sorts. And so he daringly composed his *Dialogue Concerning the Two Chief World Systems*, pushing Copernicanism and openly mocking those who resisted the new science of searching for *facts*. One middle finger, liberated, pointed to the stars.

A COUPLE OF MONTHS after my trip to Missouri, in January 2009, I drove three hours northwest from my home in East Lansing into the snow-covered woods near Traverse City, and knocked on the door of the anthropologist Napoleon Chagnon. By then I had talked to enough scientists attacked by various progressive activists that I could guess how I would find the septuagenarian Chagnon—engaged in a sort of self-imposed house arrest, treated by his peers as cancerous and contagious, portrayed by his friends as a martyr and by his enemies as a Nazi, disoriented, ineffectively angry, and essentially stuck at the moment his controversy had fully broken. I guessed about right. One might just add to the image a blue and gold University of Michigan baseball cap, an affinity for rambunctious dogs and children, a taste for bawdy jokes, and a seemingly endless thirst for coffee or beer, depending on the time of the day.

I had decided to carefully investigate Chagnon's story because his was said by scientists I now trusted to illuminate like no other the dangerous intellectual rot occurring within certain branches of academe—the privileging of politics over evidence. Chagnon's appeared to be a story of what happens when liberal hearts bleed so much that brains stop getting enough oxygen. Although I had no hope of curing this pathology now infecting parts of the ivory tower, I thought it might at least be useful to study and describe an index case.

Long before accusations against him started making front-page news, Napoleon Chagnon had gained worldwide renown for his groundbreaking studies of the Yanomamö, an indigenous people who live in sparsely populated rain forest where Brazil meets Venezuela. Besides living among the Yanomamö and learning their language, myths, and rituals, in the 1960s Chagnon began gathering gigantic amounts of data on Yanomamö genealogies, movement of villages, gardening and hunting practices, infanticide, nutrition, causes of mortality, and on and on. While many of his colleagues in cultural anthropology were collecting and producing largely qualitative data—*stories* of various peoples—Chagnon wanted to make his study as aggressively scientific as possible. He seized all available opportunities to study the Yanomamö *quantitatively,* looking precisely, for example, at how causes of death correlated with age and sex, at protein intake, and at kinship patterns. Indeed, Chagnon was so oriented to the quantitative that he was one of the first anthropologists to bring a computer to a remote field site. This extraordinarily deep and broad work on a relatively isolated indigenous people was a boon for science.

From the 1970s to the 1990s, pretty much every American college student who took an introductory anthropology course learned some part of Chagnon's work. Many other Americans came to know his work via popular magazines like *National Geographic.* But Chagnon's growing public fame had been steadily matched by growing infamy within his own field. That was in part because Chagnon had been an early and boisterous defender of sociobiology, the science of understanding the evolutionary bases for behaviors and cultures. Even so, by Chagnon's time, all anthropologists believed in human evolution, and so his interest in studying humans as evolved animals might never have gotten him in so much trouble were it not for a couple of other things.

First, Chagnon saw and represented in the Yanomamö a somewhat shocking image of evolved "human nature"—one featuring males fighting violently over fertile females, domestic brutality, ritualized drug use, and ecological indifference. Not your standard liberal image of the unjustly oppressed, naturally peaceful, environmentally gentle rain-forest Indian family. Not the kind of image that will win you friends among those cultural anthropologists who see themselves primarily as defenders of the oppressed subjects they study, especially if you're suggesting, as Chagnon was, that the Yanomamö showed us *our* human nature.

So Chagnon's characterization of the Yanomamö as "the fierce people" alone could have gained him a fair number of academic enemies within anthropology, especially as cultural anthropologists moved en masse into political advocacy, including feminist and antiwar endeavors. But exacerbating tensions was the fact that Chagnon had the classic Galilean personality, complete with political tone dumbness—that inability (or constitutional unwillingness) to sing in tune. Indeed, descriptions of Chagnon provided to me by both his friends and enemies sounded eerily reminiscent of Galileo: a risk-taker, a loyal friend, a scientist obsessed with quantitative description, a brazen challenger to orthodoxy. When I came along, there remained an open question among Chagnon's colleagues as to whether he was such a rough character because he spent his formative adult years among the Yanomamö, or whether he was able to study the Yanomamö precisely because he was *already* such a tough character. (Most colleagues guessed the latter.) But no one who knew Chagnon personally imagined him to be the kind of guy who could ever have had a polite academic conversation with colleagues whose political sensibilities caused them to challenge Chagnon's view of the world.

The battles within anthropology between Chagnon and his detractors had finally exploded onto the world scene when, in 2000, following up on Chagnon's disciplinary critics, the self-styled "journalist" Patrick Tierney started publicly alleging that, beginning in the 1960s, Chagnon had committed one atrocity after another against the Yanomamö. Tierney called his book on the subject *Darkness in El Dorado: How Scientists and Journalists Devastated the Amazon.* The work focused chiefly on Chagnon and James V. Neel Sr., a famous physician-geneticist who had collaborated with Chagnon in South American fieldwork. Neel, who many people knew had been suffering from terminal cancer, died just before Tierney's work came out, and people said the timing of Tierney's publications wasn't a coincidence; a dead man can't sue for libel. But Chagnon was alive, and Tierney's claims made his life a living hell, largely because of the decision by Chagnon's colleagues in the leadership of the American Anthropological Association (AAA) to take Tierney's book and run with it.

The whole thing really took off just before Tierney's work was published. In September 2000, two anthropologists who had long been on Chagnon's case—Terence Turner and Leslie Sponsel—wrote a letter to the heads of the AAA alerting them to Tierney's soon-to-be-published work, summarizing the charges made against Chagnon and Neel, and sprinkling on lots of rhetorical pepper, including even an allusion to the Nazi scientist Josef Mengele. The Turner-Sponsel letter opened by announcing, "In its scale, ramifications, and sheer criminality and corruption, [the scandal] is unparalleled in the history of Anthropology." Turner and Sponsel then recounted Tierney's most sensational claims, including that Neel had, "in all probability deliberately caused" an outbreak of measles in 1968 by knowingly using a bad vaccine among the Yanomamö to test an "extreme" and "fascistic" eugenic theory. Turner and Sponsel accused Chagnon of supporting Neel's efforts by carrying out research that "formed integral parts of this massive, and massively fatal, human experiment." Additional charges included "cooking and re-cooking" data, intentionally starting wars, aiding "sinister politicians" and illegal gold miners, and—perhaps the most inflammatory claim—purposefully withholding medical care while experimental subjects died from the allegedly vaccine-induced measles. This stuff made the charges against Bailey look like a schoolyard brawl, especially because those against Chagnon were coming from scholars in his own field.

Before the Internet, cooler heads might have prevailed. (Insiders knew this was hardly Turner and Sponsel's first attempt to pick at the big dog Chagnon.) Instead Turner and Sponsel's juicy "tell all" letter wound up circulated all over the world virtually overnight, and of course the press didn't dare sit on such a hot story long enough to find out what was true, much less learn the backstory. Most reporters simply reiterated the charges. A headline in *The Guardian* screamed, SCIENTIST "KILLED AMAZON INDIANS TO TEST RACE THEORY." The quotation marks likely would be lost on much of the public.

At this point, Tierney's book wasn't even out yet, nor was the *New Yorker* article he'd written summarizing his most horrifying claims, but thanks to press attention to the Turner-Sponsel memo, all hell broke loose. The AAA leadership decided to convene a highly publicized special session for the upcoming meeting in November 2000 in San Francisco, and the circus quickly grew so large that it started to require extra tents. Then the AAA leadership upped the ante, creating an El Dorado Task Force to look more deeply into Tierney's claims. Although Chagnon was obviously being put on trial at the AAA, no one from the association ever issued him a formal invitation to defend himself. He was to be tried in absentia.

In sharp contrast to the AAA, various other scholarly bodies rose up immediately to object to what they saw as obvious falsehoods in Tierney's work and by implication in the Turner-Sponsel memo. Fact-based criticisms of Tierney were issued by the National Academy of Sciences, the American Society of Human Genetics, the International Genetic Epidemiology Society, and the Society for Visual Anthropology. The University of Michigan—where Chagnon had been a graduate student and then faculty member, and where Neel had done most of his work—also issued a devastating point-by-point rebuttal of Tierney's most problematic claims.

Why was the response of the AAA so anomalous? The answer can't be that the AAA leadership remained unaware of factual challenges to Tierney's claims, including devastating criticisms from Susan Lindee, a senior historian of science at the University of Pennsylvania who had written extensively about Neel. As soon as she got the Turner-Sponsel letter via e-mail, Lindee dropped everything except her class and ran over to Neel's archives at the American Philosophical Society to see if she had missed something major. On the contrary, she immediately found extensive evidence that Tierney had gotten many things wrong. She issued an open letter saying so, and later reported her findings in person at the AAA meeting. Lindee found clear signs that the outbreak of measles had predated Neel's arrival with the vaccines, so he could not have caused the epidemic, as Tierney, Turner, and Sponsel suggested. And although Tierney claimed Neel had tried to stop his colleagues from treating the Yanomamö so he could run a Nazi-like experiment to see who would live and who would die, Lindee found substantial evidence that Neel had done all he could to get ahead of the epidemic and save those who were already infected.

Lindee had hardly been alone in quickly and publicly presenting evidence that challenged Tierney's most shocking claims. Thomas Headland, a missionary and anthropologist with contacts in the region, gathered and presented additional evidence that the 1968 epidemic predated the expedition's arrival. Historians of science Diane Paul and John Beatty presented evidence that—contrary to the implications that Neel's funding from the Atomic Energy Commission meant he was up to an extraordinary bit of no good—about half of all federally funded American geneticists at the time had AEC funding. Lindee, Paul, and Beatty, historians with essentially no horse in this race, also challenged the portrayal of Neel as a Nazi-like eugenicist.

Meanwhile, one scholar of the Yanomamö after another showed evidence that, contrary to the claims of Tierney, Turner, and Sponsel, Chagnon had not invented the Yanomamö reputation for ferocity, fighting, and abducting women. In spite of Tierney's portrayal of Chagnon as the bringer of strife to a naturally Edenic people, various anthropologists and historians pointed to evidence for Yanomamö conflict and the kidnapping of fertile women dating back to long before Chagnon was even born.

Yet in spite of all these clear declarations that Tierney's book amounted to a house of cards, the AAA had gone full steam ahead. That meant the AAA essentially bolstered Tierney's claims against

Chagnon and the late Neel and provided PR for Tierney's book and *New Yorker* article, too. The Tierney-inspired free-for-all conducted under the auspices of the AAA enabled "scholars" to stand up at microphones and debate whether Chagnon was a "swashbuckling misogynist" and a fomenter of violence, to claim that various American and European scientists had been responsible for spreading Ebola around Africa, and to use the AAA Web site to throw up utterly undocumented charges against colleagues.

Some anthropologists did try to fight back. In 2003, Tom Gregor of Vanderbilt University and Dan Gross of the World Bank launched a referendum in the AAA explicitly criticizing Tierney's book and the AAA El Dorado Task Force (and thus implicitly criticizing Turner and Sponsel) for misrepresenting the Yanomamö measles vaccine history in such a way as to undermine ongoing vaccine campaigns that otherwise had the potential to save vulnerable people all over the world. The referendum passed by a ratio of 11 to 1. Then in 2005, Gregor and Gross put forth another referendum to withdraw the AAA's acceptance of the Task Force Report. The motion passed by a ratio of about 2.5 to 1. Impressive, particularly considering that by then a fair number of science-oriented anthropologists apparently had quit the AAA because of what its leadership had done to Chagnon, Neel, and science itself. (One of my closest friends in East Lansing, a scientific anthropologist at Michigan State, told me that he had dropped his AAA membership right after the AAA had tried Chagnon in the kangaroo court held at the San Francisco meeting.)

But those referenda, coming fairly late in the game, couldn't possibly undo the damage done to Chagnon's and Neel's reputations. Indeed, in some ways, they simply muddied the filthy waters more. When I came to the story, in spite of the AAA membership's vote four years earlier to rescind acceptance of the Task Force Report—to essentially take back any hint of a guilty verdict—the report remained up on the AAA Web site, without any attached notice of the rescission. It included a number of ruinous (and completely unsupported) claims, including the allegation that Chagnon had paid his Yanomamö subjects to kill each other.

As for Chagnon, it seemed pretty clear his career had essentially been halted by the whole mess. He was supposed to have retired with his wife Carlene to this house in the Michigan woods so he would have a place to hunt, to fish, to run his dogs, and to write his memoirs. But from what I was hearing, his memoirs seemed to be stalled. And perched on a bluff, reachable only down a long driveway, this house seemed to me less like a sanctuary than a fortress.

NOT LONG AFTER I ARRIVED, Nap Chagnon offered me a mug of coffee and a chair in his home office, and we sat down to start talking. I first reminded him of what I'd told him earlier regarding my standard interview method: We would talk, I would take notes, and then I'd return the notes to him. He could change them however he wished—add or delete anything—and I would use only what he approved as being on the record. At that, he started his story, and I started typing. Soon he paused to express skepticism that I could type fast enough to actually get down what he was saying. I read back to him exactly what he had said so far, including his skepticism about my typing speed. He raised his eyebrows and we really got down to business.

Subjects like Chagnon—people who spent their lives as professional interviewers of sources—prepare in advance what they will tell you, and there is no way to redirect them. I knew this, and so I braced myself for waiting out what it was he felt I needed to know, before we could get to what I wanted to know. Chagnon's story was by turns fascinating and complex, circular and gossipy, important and banal. I felt rather as if I was trying to drink from a fire hose, but I just kept typing and nodding, stopping him only once in a while to ask him to spell a name for me.

Chagnon's story tended toward the tribal; that is, it was pretty clear there were people on his side (good guys) and people against him (bad guys). The undercurrent included a story of social class, one that made a lot of sense to me. The physician Neel had been upper-crust and well established. He'd been a man who tended to keep his whites white even in the jungle, a man who had immodestly titled his autobiography *Physician to the Gene Pool.* Chagnon, by contrast, came from a large working-class family and had had to struggle to make his way into the world of universities. Chagnon seemed to have understood from the start that he'd never really be welcomed into the blue-blooded Ivy League anthropologists' club, no matter how important his work became. I got the sense that, down in Venezuela, Chagnon had more readily related to the Yanomamö than to some of the American academics on the expeditions. He knew what it was like to hunt for your food, to get tispy with your chums around a campfire, and to sit around telling lewd stories.

Chagnon completely immersed himself in life with the Yanomamö, quickly becoming vastly more conversant in their rich language than Neel ever would. Chagnon's 1968 monograph tells of his hardships, his adventures, and his scientific findings, portraying neither himself nor the Yanomamö as heroes or saints. In fact, Chagnon admitted in that book to having been fooled by the Yanomamö during data collection. They took advantage of his inexperience to introduce all sorts of scatological jokes into the eager young anthropologist's records, such that Chagnon lost months of work that had to be completely done over. The book tracks the lives of "the fierce people," focusing in depth on the forms of and motivations for fighting, including especially women-stealing. From it, I understood why Chagnon got pissed off when people referred to him as Neel's assistant or graduate student. He had been neither, but more to the point, he'd been pretty damned ballsy to do the work he'd done out there in the name of anthropology. And he'd brought back an astonishing store of scientific data.

As we talked, Chagnon fleshed out for me just how far back Terence Turner's obsession with him went. Tierney's book had hardly been the start of Turner's attacks; it had simply been the best ticket Turner had ever gotten to ride. Now at Cornell, Turner had been trying for more than a decade to go after Chagnon with claims of bad behavior, and to some extent, he had joined with South American anthropologists, who (Chagnon said) didn't appreciate Chagnon's competition in "their" area. They had all made it harder and harder for Chagnon to get research permits in the area, until finally he had had to give up. That led to his decision to retire from the University of California–Santa Barbara (UCSB). He wasn't interested in endless teaching and committee work without any prospect of getting back to the indigenous people he'd come to know and make known.

A few years before he left UCSB, though, Chagnon had gotten hints about what might be coming from Tierney. A book rep from Norton had stopped by during a visit to another faculty member to warn Chagnon confidentially. After talking to the rep, Chagnon wrote to Neel and another colleague who was implicated—a Venezuelan physician named Marcel Roche—to warn them. Apparently in response to Chagnon's letter, Neel pulled his relevant field notes and related materials, items that could prove conclusively what happened in 1968, and put photocopies of them all in a single folder marked YANOMAMA-1968-INSURANCE—insurance against lies. In 2000, that message-in-a-bottle from the late Neel was one of the key folders in the American Philosophical Society archives in Philadelphia that Lindee used to counter Tierney's claims.

After many hours of listening and transcribing, I told Chagnon my fingers were going to fall off if we didn't take a break, and I suggested we start back up in the morning the next day. I asked him what time he wanted to reconvene. He suggested five thirty—nearly three hours before dawn at that time of year. His wife, Carlene, objected, but I told her I had come to do this work, and if that was what time Nap wanted to start, I'd get up. But then I renegotiated for six A.M. I asked Chagnon if he had always been a morning person. He said no, it was an old habit from the field, back when the village he'd stay in was so small that as soon as light broke the "damned Indians" would wake him up with their morning noise. The way he said it suggested he missed it.

Before I went to bed, I stood in the living room for a while with Carlene, looking up at the big photographic prints on the wall. These were beautiful photos Nap had taken of the Yanomamö decades before. He was an astonishing photographer; any museum would have been glad to mount an exhibit of this work. One picture featured a beautiful little boy practicing shooting an arrow up into the air. Another featured a tender moment between a mother and child. And still another, a group of men negotiating a possible trade of a dog, the dog in the center clearly nervously aware of what was afoot.

"Tell me," Carlene said to me, her eyes starting to water, "how can they say the man who took these pictures would hurt these people? How can they say that?" It was obvious to her, as it was now to me, that these were essentially family photos. People like Tierney, Turner, and Sponsel seemed to want to make a special claim of being the defenders of oppressed people like the Yanomamö, to position themselves as the white hats to Chagnon's black. But Chagnon seemed to have a gut-level sympathy for the Yanomamö, a sympathy perhaps less articulate than his critics', yet easily as deep. Claims that Chagnon's work had harmed the Yanomamö were the ones that stung him most sharply. And the truth was that, in practice, he had actively tried to stop use of his data to oppress the Yanomamö; for example, when he found out that the data he had collected on Yanomamö infanticide might be used by the Venezuelan government against them, he had essentially withdrawn the data. Like Bailey, like Palmer, like so many others, this was a scientist out primarily for truth, but never at the *cost* of justice.

The next day, in the dull dark of the winter Michigan morning, we started again, Chagnon offering me only black coffee, Carlene still in bed, their hunting dog Darwin, a German shorthaired pointer, lying on the floor next to us, alternately farting and snoring. I started to walk Chagnon through what I really wanted to know—what wasn't already in the record. What had it been like, surviving this controversy? What moments came back to him when he thought about living through it?

He told me he remembered, particularly vividly, how much it had meant to him when he got the call from Danny Gross telling him of the successful vote among the AAA membership to rescind acceptance of the AAA's El Dorado Task Force Report. He told me of his colleagues at UCSB—Ed Hagen, Michael Price, and John Tooby in particular—fighting back by showing Tierney's twisted use of sources and quotations. And he told me of how his friend Ed (E. O.) Wilson at Harvard, the founder of sociobiology, had called him every week to make sure he knew he was not alone.

Decades before, Chagnon had been the one defending Wilson. The most vivid instance had to be the time in 1978 when Wilson was presenting about sociobiology at a special session of the American Association for the Advancement of Science in Washington, D.C. Wilson had broken his leg recently, so he was in a cast that stretched almost from his ankle to his hip. As he tried to speak, several members of the self-proclaimed "International Committee Against Racism" had rushed the stage, declared Wilson all wet, and dumped a pitcher of water over his head. Chagnon had been at the back of the auditorium, but he rushed up to try to knock heads together as necessary. Probably best for all, the ensuing pandemonium blocked Chagnon's path. But Chagnon had been more effectual in various other venues at defending Wilson, particularly against what Wilson felt were misrepresentations by Richard Lewontin and Stephen Jay Gould, who were also at Harvard and who claimed to speak for the oppressed.

When the El Dorado storm hit, Wilson made sure to call Chagnon often to remind him of what mattered and give him some sympathy. As he told me of the colleagues who had moved to help him, Chagnon started to choke up. Naturally, I started to lose it too, as I always do in such situations, but I just let the tears dribble down my cheeks without making a noise. I typed as quietly as I could. Then Chagnon stood up suddenly and announced to me and Darwin, "I have to go to the *men's* room." He climbed around Darwin and made his way out of his home office to the hallway bathroom.

When he returned a few minutes later, he started in on a completely different story, confusing me thoroughly. It was a story about driving around Washington, D.C., with Margaret Mead. I knew from what Chagnon and others had told me that he and Mead had had a long and probably somewhat contentious scholarly relationship, given their very different portraits of human sexuality. In her popular 1928 book, *Coming of Age in Samoa*, Mead had presented Samoan culture as one that, without much fuss or fear, allowed many adolescents to sexually experiment—to fool around in ways that in the 1920s would have shocked the average American. This was a very different story of "primitive" human sexuality from Chagnon's, which told of Yanomamö males regularly engaging in fierce fighting over women. By Chagnon's account, human sexual relations tended to be tense affairs, involving violent abductions and even homicide. By Mead's, human sexual relations meant making love, not war. Moreover, Chagnon had leaned toward biological explanations for what he saw as commonalities in human behavior, whereas Mead was more inclined to notice cross-cultural *differences* and explain them via social structures.

Whatever their dissimilarities in worldview, however, Mead had always been a mensch for science—for free inquiry and free speech—and Chagnon appreciated that. He hastened to tell me of the time he and his best friend, the anthropologist Bill Irons, had tried to hold the first session on sociobiology in anthropology at a AAA meeting in the 1970s. A motion had been put forth to cancel the session because of the supposed dangers of sociobiology. Mead stood up and said the attempted ban was akin to book burning. Her words turned the vote in Chagnon and Irons's favor, saving the session just moments before Ed Wilson arrived to participate.

Now, back from his temporary retreat to "the men's room," Chagnon wanted to tell me another story of Mead as mensch. When the politicking against him in South America got so bad that Chagnon was being denied access to the field, Mead offered to go with him to the Venezuelan embassy in Washington, D.C., to try to fix the situation. Driving the two of them to the embassy, Chagnon became hopelessly lost in the diagonal-and-circular maze of Washington streets. He recalled to me, with his face downcast, that Mead had harrumphed at him: "You would think this famous anthropologist who can find his way all around the jungle could find his way around a city in the U.S.!"

Chagnon paused and looked up at me. "You would think this mean, nasty anthropologist could hold it together when being interviewed."

I stopped typing and said to him, in as manly a fashion as I muster, that this shit was hard on a person, and it was understandable someone might get a little emotional.

He seemed at that moment a strikingly ordinary old man—wizened, mortal, spent.

MARGARET MEAD DIDN'T LIVE to see the ruining of her professional reputation. The New Zealand anthropologist Derek Freeman didn't publish his book *Margaret Mead and Samoa: The Making and Unmaking of an Anthropological Myth* until 1983, five years after Mead's death. After that highly publicized work from Harvard University Press, Freeman's misleading claims about Mead went through even better publicized iterations, and with each pass, they had more successfully damned Mead's scientific reputation. By 1999, with the publication of *The Fateful Hoaxing of Margaret Mead*, Freeman had refined and simplified the story, marketing it perfectly for the sound-bite world: Determined by her personal political agenda to paint a sex-positive view of Samoan adolescence, Mead had allowed herself to be duped by two young Samoan women who had simply been joshing her with sexual fish tales of licentious adolescence. Mead had completely misrepresented human sexuality to the world because she'd been stupid enough to buy the joke these two young women were playing on her. Just an ideologically blinded dupe, according to Freeman, Mead turned out to be a dangerous spoiler of the scientific record on the nature of sex. In

Freeman's words, "Never can giggly fibs have had such far-reaching consequences in the groves of Academe."

To pull off the fiction he had spun as nonfiction, Freeman had employed a rather brilliant methodology. First, he made sure his work appeared on the surface to be pure scholarship—just one expert anthropologist using good data to dismantle the supposedly shoddy data collection of a predecessor. Although it was riddled with misrepresentations of Mead, Freeman's "scholarship" looked real enough to pass (at least until good scholars came along to check it).

Alongside his dishonest rewriting of Mead's data, Freeman also rewrote Mead's worldview, making her out to be an absolute cultural determinist hostile to any biological or evolutionary explanations of human nature. By portraying her erroneously as so scientifically simple-minded and so obviously outdated, Freeman was able to persuade people to write Mead off—and not only folks at the biological-determinist (sociobiological) extreme of the nature-nurture controversy, but also people in the theoretical middle. After all, what kind of idiot thinks we're all nurture and no nature!

In reality, Mead, like most twentieth-century scientists, had a reasonably complex view of human behavior, assuming and seeing contributions from both biology and culture. That was one reason she was sympathetic to the work Chagnon was doing; he, too, saw both biology and culture as important. But Freeman needed Mead to be an extremist as well as a dupe, a kind of groovy 1960s antiscience anthropological tourist in a big straw hat.

This image then allowed Freeman to deploy the last bit of his clever methodology—to swoop in to play, in his own grandiose words, "The Heretic" to the supposed Cultural Church of Margaret Mead, to make it a battle between two "greats," thus making himself as great as Mead. By boldly reducing Mead to a big-name ideological hack, Freeman could play Galileo, saving science from mere dogma. Freeman appeared to be not only a brilliant scholar, but a hero, as well.

Freeman was well into his relentless assault on Mead's reputation when a number of cultural anthropologists tried to step in to right the factual wrongs. In his 1996 book, *Not Even Wrong: Margaret Mead, Derek Freeman, and the Samoans*, the anthropologist Martin Orans used Mead's own field notes to show "that such humorous fibbing could not be the basis of Mead's understanding. Freeman asks us to imagine that the joking of two women, pinching each other as they put Mead on about their sexuality and that of adolescents, was of more significance than the detailed information she had collected throughout her fieldwork." Freeman had thoroughly misrepresented Mead's work, creating such a fantastical account that it was "not even wrong" because it was essentially fiction.

Finding himself as appalled as Orans had been by Freeman's defamation of Mead, the anthropologist Paul Shankman of the University of Colorado–Boulder decided to devote a sizable chunk of his professional energies trying "to extricate Mead's reputation from the quicksand of controversy." Performing expert historical analysis, Shankman showed that what Mead *really* drew on for her conclusions was data "collected on 25 adolescent girls of whom over 40% were sexually active" and from interviews with Samoan men and women. In his 2009 book, *The Trashing of Margaret Mead: Anatomy of an Anthropological Controversy*, Shankman readily acknowledged that Mead saw herself as "a citizen-scientist. Not content with being a bookish academic, she wanted to be a public intellectual and activist, using ethnographic data to address important public issues." In other words, yes, she had a political agenda. In writing up her populist gospel of sexual permissiveness (a gospel she lived in her personal life), Mead certainly had oversimplified Samoan society, downplaying violent rape and the fact that women were discouraged from reporting, and downplaying, too, the beatings delivered upon those who violated sexual norms. But following Orans, Shankman was able to use Freeman's *own records* to show that he *knew* Mead's work to be substantially more sophisticated and rigorous than his negative portrait of it. In Shankman's words, "Freeman was able to advance his argument only by very selective use of information, including the creative use of partial quotations and the strategic omission of relevant data at crucial junctures in his argument."

Curiously, when he died in 2001, Freeman intentionally left behind in archives the documentation that would ultimately undo him. Perhaps he had so completely bought into the tale he had spun that it never occurred to him to fear leaving this self-defeating evidentiary trail. As Shankman discovered, the stash includes a key interview with one of the two "joshing" informants, showing that the interview was set up to get the informant to turn on Mead. Eerily reminiscent of the twisted facts in the Bailey controversy, in his analysis of the Freeman-Mead controversy, Shankman also found that "there is no information on sex from these two women in Mead's field notes." In other words, Freeman misrepresented not only the role of Mead, but also that of the two supposed "joshing informants."

But why? Shankman's digging suggests that not only was Freeman interested in pushing a particular biology-heavy view of human nature (for which he needed a naive Mead as foil), but he also suffered from a weird obsession with Mead, whom he saw as a threat to anthropology, to Samoans, and even to himself. He appeared to genuinely believe it was his duty to be the big man who would take Mead down. Freeman had no patience for those who might get in his way and sometimes threatened those who did. Perhaps even more disturbing, Freeman seems to have shown signs of delusion—*deep* delusion. Shankman tells of Freeman's 1961 trip to Sarawak, where he viewed sexually graphic tribal statues at a local museum: "Freeman was convinced that the erotic statues [there] not only were a perversion of authentic tribal culture but were also exerting a form of mind control over Freeman through their hypnotic power, a power that he was determined to break. Freeman also believed that the statues were being used by [a colleague] and the Soviet Union to subvert the local government. Indeed, Freeman thought [the colleague's wife] was a Soviet agent." Freeman then proceeded to destroy one of the statues. Concerned by his bizarre behavior, authorities eventually banned Freeman from further research in Sarawak. One of his colleagues told Shankman, "We all know he's crazy, but we can't say it!"

So as it turns out, it was not Margaret Mead or her supposedly joking informants but the strange Derek Freeman who managed to "hoax" the world. Freeman succeeded in part because he followed what I had learned is the number-one rule in making shit up: Make it so unbelievable that people have to believe it.

CHAPTER 6

HUMAN NATURES

ALTHOUGH *DARKNESS IN EL Dorado* made Patrick Tierney look like an extremely adventurous but scholarly investigative reporter, in fact Tierney had no apparent training or employment history in anthropology or journalism. His first book, *The Highest Altar,* had purported to reveal ongoing human sacrifice in the Andes. No one in the scholarly world appeared to take that book all that seriously. But *Darkness in El Dorado* was a very different sort of book, absolutely crammed with impressive-looking footnotes—so many that the book looked like a masterwork of objective scholarship.

When she interviewed Tierney about *Darkness in El Dorado* in late 2000 for Chicago Public Radio, Victoria Lautman made specific mention of Tierney's apparent documentation of his claims:

> There are 60 pages just of footnotes supporting Tierney's incendiary main point[s], namely that the Brazilian Yanomamö Indians were hideously exploited, that a lethal 1968 measles epidemic was spread by a dangerous vaccine, that the U.S. Atomic Energy Commission used the Yanomamö as a control group without their knowledge, and, most important, that all of these shocking abuses were perpetuated by two of the most famous and respected members of the anthropological community [*sic*].

As I looked back at all the positive media attention and praise the book got—it had even been named a finalist for a National Book Award—there could be no question this had resulted from readers assuming the footnotes were real. The truth was that plenty of them had simply not checked out. I knew this from reviewing the work of previous scholars who had looked, but I also was finding still more examples on my own.

For instance, as I went over the *Darkness* chapter on the 1968 epidemic, I came across this line: "The vaccinators were Napoleon Chagnon and a respected Venezuelan doctor named Marcel Roche." Chagnon had told me repeatedly that he had not vaccinated anyone during the epidemic. This point mattered a lot to him, because Tierney's *New Yorker* article included a story of a man whose child had allegedly died following a vaccination from Chagnon. Chagnon was understandably distraught at the implication that he had killed a Yanomamö child. So I looked at Tierney's citation for the claim that Chagnon and Roche had been vaccinating and was rather stunned to see that Tierney seemed, by the citation, to be attributing this information to an article Chagnon had coauthored in 1970. How could Chagnon tell me he didn't vaccinate anyone during the epidemic when his own coauthored article said that he did?

I wasn't looking forward to having to confront the scarred and forceful Chagnon with that question, but I knew I had to. So I pulled the 1970 article and went to page 421, as Tierney's citation indicated I should. Nowhere on the page did it name any vaccinator. Confused, I went through the rest of the article. *Nowhere in the article was a single vaccinator named.* Tierney's citation was full of gas.

As I moved through what would become a year of research and about forty interviews for this project, a clear pattern of misrepresentation emerged. Even people who had been relatively aligned with Tierney were now admitting to me that he had played fast and loose with the truth. I called to interview Brian Ferguson, an anthropologist at Rutgers who had written *Yanomami Warfare*, a book highly critical of

Chagnon. In the book, Ferguson argues that the introduction of large amounts of Western trade goods by researchers and missionaries contributed to Yanomamö conflict that Chagnon often blamed on sexual tensions. Ferguson told me that when the *New Yorker* fact checkers called him,

> everything was fine except one passage where Tierney has me saying something to the effect of "missions could be disruptive but according to Ferguson they are less so than Chagnon was," downplaying the impact of the missions. I said, no I didn't say that, and I don't believe that to be true. I think [the missions] were very disruptive in the period I'm talking about. . . . I said that's not what I said. And I got a call from Patrick Tierney and he got quite angry about it and said that I was backing down and that I was making a political move here and that he had me on tape saying what he said I said. And I said you'd better get that tape ready, because that's not what I said.

Another strike against Tierney came from a woman with whom he'd apparently had a close friendship in South America, a woman named Lêda Martins, now an anthropologist at Pitzer College in California. Before going into anthropology, Martins had been a journalist and human rights worker, and she had long shared Tierney's concern for the indigenous peoples in Venezuela and Brazil. In the acknowledgments to *Darkness*, Tierney said he was "especially indebted" to Martins, adding, "Leda's dossier on Napoleon Chagnon was an important resource for my research." I knew this dossier to be very important—Chagnon was practically obsessed with it—because it contained many of the misrepresentations of *Darkness* yet predated the book by years. Indeed, the copy of the dossier that Chagnon had obtained and given to me read almost like a draft book proposal for *Darkness in El Dorado*. The dossier had been used in various ways, but mostly to try to get

Chagnon's research permits denied. It was probably largely responsible for forcing an unwanted end to his fieldwork.

So, in his book's acknowledgment, Tierney basically was saying that Martins had established many of the most damning charges against Chagnon. Martins's charges against Chagnon would then constitute the basis for what Tierney would have followed up. I pressed Martins for a copy of the dossier as she had it; I wanted to know if it was the same as the one Chagnon had gotten his hands on. Eventually, when I went to meet her in person while I was in Southern California, she handed me a copy. It turned out to match Chagnon's copy. But at that time, she also confessed something key, something she later, at my request, confirmed in an e-mail. She was not the author of the dossier. In fact, Martins told me:

> Patrick Tierney wrote the Chagnon dossier and I translated [it] to Portuguese. . . . I presented the dossier to Brazilian authorities (Funai employees) and human rights advocates who were looking for information on Chagnon who was seeking permission to go inside the Yanomami Territory in Brazil. I was the one who circulated the dossier in Brazil because people knew and trusted me. I trusted Patrick and did not check his references. (I can only hope whatever is left of my friendship with Patrick will survive the truth, but . . . he should not have said that.)

So the truth was that Tierney himself had written the charges he attributed to Martins, and Martins, presuming them to be true, had used them against Chagnon. This meant that Tierney had been working to spoil Chagnon's reputation and his ability to do fieldwork in the Amazon many years before *Darkness in El Dorado*.

Chagnon had long suspected that the dossier charges originated with Patrick Tierney and that he had long been near the center of Chagnon's troubles, even though in *Darkness* Tierney looked like an objective reporter, not a human rights activist. After I interviewed Chagnon's longtime collaborator Raymond Hames of the University of Nebraska, Hames dug up for me a remarkable e-mail that he'd received from Chagnon on November 6, 1995, five years before the publication of *Darkness in El Dorado*. Chagnon had written to Hames:

> I finally made, with the help of a Brazilian friend, a translation of the "Dossier" on me that is circulating in Brazil and was used in September to try to have FUNAI rescind my [anthropological research] permit. It is so hysterical and preposterous that it is funny, but there will be lots of people who will believe the[se] claims. . . . Footnote #21 leads me to suspect that the primary author of this is one Patrick Tierney, who actually showed up in my office just after I returned from Brazil. I pointedly asked him if he were aware of this "dossier" and he denied any knowledge of it. I think he is a liar.

It sure seemed that Chagnon was right to be deeply suspicious of Tierney. When I started my research on all this, I thought perhaps Tierney had just been sloppy here and there, that perhaps he had committed wishful thinking in various places and accidentally misordered events. But so much of what he put forth turned out to be inaccurate. I found myself mulling the strong claim made in 2001 by a group of Chagnon's colleagues at UCSB who had looked into the matter: "The major allegations against Napoleon Chagnon and James Neel presented in *Darkness in El Dorado* by Patrick Tierney appear to be *deliberately* fraudulent."

The Bailey history had impressed on me the importance of understanding the backstory of a controversial book, particularly the relationships behind the text. So I knew I needed to try to understand the prepublication relationships of those involved in the *Darkness in El Dorado* controversy. Deploying what Aron liked to call the historian's secret weapon—um, a timeline—I was able to see that right around the time Chagnon had been e-mailing his colleague about his suspicion that Tierney had authored the dossier—right around the time that Chagnon had been guessing Tierney was "a liar"—Patrick Tierney had been introducing Lêda Martins to Terence Turner, Chagnon's longtime nemesis, at the Pittsburgh airport.

Why were these three—the writer Patrick Tierney, the activist-journalist Lêda Martins, and the anthropologist Terence Turner—meeting at the Pittsburgh airport? In the mid-1990s, back in the pre-9/11 days when people without tickets could hang out with you at the gate, Turner was regularly making flight connections there, commuting from where he worked (Chicago) to where his wife worked (Ithaca, New York). Martins had found herself in Pittsburgh on a Fulbright to learn English. And Tierney's family was based in Pittsburgh; his father was a professor at the University of Pittsburgh. Turner acknowledged to me that the three of them had met at the Pittsburgh airport several times. And around the time of these meetings came the fieldwork-ending dossier, after them came *Darkness*, and then Martins decided to go on to get a PhD in anthropology . . . and earned it under Terence Turner. She used that PhD in part to go after Chagnon.

The trio of Turner, Tierney, and Martins had been quite effective in clouding Chagnon's reputation. In 2001, for example, when she was supporting the AAA-centered prosecution of Chagnon, Martins had publicly taken Chagnon to task for what he had said to a popular Brazilian magazine regarding the Yanomamö. To the magazine reporter, Chagnon had observed, "Real Indians sweat, they smell bad, they take hallucinogenic drugs, they belch after they eat, they covet and at times steal their neighbor's wife, they fornicate, and they make war." Martins used this quotation to try to show that Chagnon dehumanized the Yanomamö and so allegedly threatened their well-being. However, when she reproduced this passage, she didn't mention how Chagnon concluded his statement to the magazine: "They are normal human beings. This is reason enough for them to deserve care and attention."

Among these three—Martins, Tierney, and Turner—Turner certainly could boast the oldest and most persistent pursuit of Chagnon in the name of ethics. *Why,* Chagnon kept asking me, *why* had Turner had this unrelenting obsession with him that lasted for so many years and that seemed to sweep others in? Why did he seem to sic every dog on Chagnon? I didn't know and couldn't know. Besides, the most interesting question to me was not that one but rather how Turner had managed to push the AAA to do what it did to Chagnon and Neel, against the example of every other scholarly organization involved. The more I dug, the more the answer to my question seemed to be Tierney's objective-looking book, *Darkness in El Dorado.* Most outsiders would know nothing of the complicated, sometimes (as in the case of the dossier authorship) actively obscured relationships among Tierney, Turner, and Martins. Outsiders wouldn't know that Tierney's source use looked much more real than it actually was. *Darkness* looked to most readers like an objective, scholarly confirmation of apparent strangers' long-running suspicions that Chagnon and Neel had committed deeply unethical acts.

Of course, the AAA might have done what the other groups did and scrutinize Tierney first, discovering the inaccuracies and maybe also the complicated relationships. But goaded by Turner and Sponsel and their concerns about egregious human rights violations, they instead trundled on, "giving voice" to what turned out to be baseless accusations allegedly made on behalf of oppressed indigenous peoples. And so the AAA bolstered Tierney's work, helping it grow legs it never could have otherwise.

Chagnon understood Turner's hands to have been stained in this, but he was also convinced that the Catholic Church had a lot to do with it all, specifically the Salesian missionaries, with whom he had come to blows. Chagnon had published work arguing that the Salesian missions could be damaging to the Yanomamö, and so they'd been fighting for years. During our interviews, Chagnon told me repeatedly that, in the early 1990s, someone had started distributing anonymous packets condemning him for various alleged offenses against the Yanomamö. These had been mailed to Chagnon's colleagues around the country and even mysteriously showed up in stacks on handout tables at an AAA conference. In my digging, I came upon an article providing evidence that Chagnon's suspicion was right: These had been distributed by the Salesians.

Even after finding that evidence, I thought a story prominently featuring a kind of Catholic persecution of Chagnon couldn't be true—it seemed to match too conveniently the story of Galileo. But the more I looked into Tierney, the more it seemed Chagnon might be on to something. Tierney certainly didn't come across to me as a fervent Catholic in *Darkness in El Dorado.* But then on a tip from an anthropologist and with the help of good librarians, I got my hands on a copy of an unpublished book by Tierney that showed him in a completely different light. This was meant to be Tierney's second book. It was supposed to have been published by Viking sometime around 1994 under the title "Last Tribes of El Dorado: The Gold Wars in the Amazon Rain Forest." For some reason, it never appeared, and Tierney's second published book turned out to be *Darkness in El Dorado.*

I tried and tried to get someone at Viking to tell me why "Last Tribes" reached such a late stage—the copy I obtained from a library in Wisconsin was printed and bound, complete with a professionally designed cover and advance reviews—but was never actually published. Viking wouldn't give. All I could learn came from the book itself. I realized, as I read its four hundred pages, that I was probably one of only a handful of people who had ever bothered to find and read this book. But I wasn't reading it for information on rain forest gold mining. I was reading it as the AAA should have read it—to learn about Patrick Tierney, the man they chose to follow into a full investigation of Chagnon. And boy, did I learn.

Like Tierney's first book (*Highest Altar*), "Last Tribes" tells a first-person story of the author's explorations in South America, in this case following Tierney's earnest attempts to gather dirt on the illegal gold-mining operations that are harming the rain forest and the native peoples, including the Yanomamö. In "Last Tribes," Tierney—who, mind you, in *Darkness in El Dorado,* moralizes on every possible point about Neel and Chagnon's supposedly unethical behaviors in the field—admits to having repeatedly lied about his own identity, even to indigenous people, ostensibly to further his activist journalism. He reveals that he faked identity documents to pass himself off as a Chilean gold miner and that, as part of his disguise, he carried mercury into the rain forest, even though he knew perfectly well the devastating effects of mercury on the habitat. "Last Tribes" also shows that Tierney illegally purchased a shotgun and carried it into the indigenous territories; wandered into remote villages without first undergoing appropriate quarantine; and trekked into Yanomamö lands without first obtaining the required legal permission from FUNAI (Fundação Nacional do Índio, the National Indian Foundation), the agency charged with this travel regulation. (FUNAI was the agency to which Tierney delivered his dossier in an attempt to stop Chagnon from doing further research with the Yanomamö.) Heck, in "Last Tribes," Tierney even brags about having met up with self-confessed murderers—men he knew were wanted by the police—with apparently no thought of reporting their whereabouts to the police. Perhaps to prove how important he has managed to make himself in his muckraking, he also reports that he seemed to have gotten another man killed because the other guy was mistaken for the troublemaking Tierney.

Even more telling, "Last Tribes" reveals that, during this "work," Tierney was housed, fed, protected, and encouraged by local Roman Catholic priests. When going into remote regions disguised as a gold miner, Tierney left his money and real identity papers with the priests. When he got arrested, it was Bishop Aldo Mangiano who sprung Tierney from jail. All this was pretty much confirmed for me in a later interview in Cleveland with Father Giovanni Saffirio, a Consolata missionary priest and anthropologist who had known both Chagnon and Tierney through their various interests in Venezuela. Chagnon had for a time been an advisor to Saffirio when Saffirio was earning his degree in anthropology. When I asked him to tell me about Tierney, Father Saffirio responded in his heavily Italian accented English that Patrick Tierney was "a good Catholic" and said that Bishop Mangiano "was pleased to help him."

When I made the arrangement to travel to Cleveland to see Father Saffirio, I wasn't sure that the trip would be worth it. But it was. As we talked in the rectory of Saint John Nepomucene, a church where he was staying for a time, Saffirio told me long and interesting stories about the missions in the area of the Yanomamö, about how they sometimes really hurt people as they intended to help them. The old priest clearly appreciated the complexity of human beings and yet he knew how to balance judgment with forgiveness. I got the sense from him that Tierney had started off doing activism and advocacy that might have really been helping.

"Bishop Aldo Mangiano found Tierney helpful because he was telling American and European people what was really happening in the forest," Saffirio told me. He meant that Tierney was working to expose the horrific effects of mining—destruction of habitats, of indigenous homelands, and so of entire cultures. But Tierney had obviously used some problematic techniques; Saffirio told me that Tierney had "traveled through the area [where the Yanomamö live] with a fake ID card saying he owned a mining company in Chile where he was born. He cheated gold buyers saying he was eager to open a *garimpo* (a mine) in Roraima." Whatever Saffirio thought of this particular charade, he clearly felt that *Darkness in El Dorado* amounted to a great injustice to Chagnon and Neel.

"Chagnon is a great scholar," Saffirio said to me, adding, "He didn't make up stuff. The data he gathered were done properly." He went on:

> Whatever his personality is, he did a great scholar[ly] job among the Yanomami. For one, it was Chagnon that made the Yanomami known worldwide with books, hundreds of

> articles and dozens of documentaries, inspiring many anthropologists and scientists to do research among them. When Tierney writes negatively about Chagnon, it hurts me because Chagnon helped the Yanomami in his own way. Sure, by doing research among the Yanomami he earned a lot of money and fame, he drinks beer, at times his temper can be short, but what that matters [*sic*] in the big picture of a fine scholar?

From Cleveland I drove to Pittsburgh to interview the anthropologist John Frechione. I was under the impression that Frechione had helped to provide a visiting scholar appointment for Tierney at the University of Pittsburgh, one that appeared to still be active, and I wanted to know about that and whatever else Frechione wanted to tell me. This university appointment was surely one of the accouterments that had allowed Tierney to look like a real scholar. Frechione informed me as we started talking that since *Darkness in El Dorado,* he and Tierney had been collaborating on a project aimed at showing the supposed remaining ethical problems with Neel's behaviors during the 1968 epidemic. I listened tensely and said I'd be happy to look at the evidence they claimed to have, but mostly I was stunned that these people were still at it so many years after the AAA debacle. Was Pittsburgh some kind of anthropological zombieland, where the deadest of claims kept rising?

I knew that, after Tierney's work had emerged in 2000 and was being shot full of holes, Frechione had given Tierney a hand by doing an important interview with a physician named Brandon Centerwall. Brandon was the son of the late physician Willard "Bill" Centerwall, another American who had been on the ill-fated 1968 expedition. When he'd returned, Bill had told his teenage son Brandon about the epidemic. Bill made the claim to Brandon that Neel had wanted to let the epidemic run unchecked in the village of Patanowa-teri to see what would happen to the vulnerable Yanomamö populace. By doing so,

Neel would have been unethically testing quasi-eugenic theories of fitness. Bill claimed to his son that he had stood up to Neel, calling him to a higher moral standard, and that Neel had given in and treated the ill. This story—recorded in a 2001 interview with Brandon Centerwall by Frechione to help Tierney defend his book—seemed proof positive that Tierney, Turner, and Sponsel were right: Neel had been a heartless eugenicist who had let infected Yanomamö die during the 1968 measles epidemic.

For a long while, the Centerwall story had been like a peppercorn stuck in my molar. All the documentary evidence suggested it couldn't have happened the way Bill Centerwall had told his son, with Neel supposedly wanting to let the epidemic run unchecked. But both Frechione and Turner had Brandon *on record* remembering his father's vivid story of Neel's cold-blooded plan and Bill Centerwall's lifesaving intervention. So I did what I had to do: I asked the son to try to explain the discrepancy between his story and the rest of the historical record.

In my digging to find him, I discovered that Brandon himself had been at the center of a little controversy; a student of literature as well as a physician, Brandon Centerwall had apparently written a dangerously persuasive scholarly article suggesting that Humbert Humbert, the pedophile of Vladimir Nabokov's *Lolita,* actually was a veiled self-portrait of the author. I hoped surviving a controversy might make Brandon more inclined to speak with me, not less. I felt strange having to question a man's late father's story, but it obviously mattered rather a lot. After confirming by e-mail that I had the right Brandon Centerwall, I wrote to ask him to confirm Frechione's and Turner's accounts of his retelling of his father's story.

No answer.

I wrote again five days later. Would he talk to me?

Two days later, Brandon answered, but cryptically, in a short message saying that he would send me a real reply later that day. He added, "I give you full permission to do as you wish with it, including quoting any or all of it, or sharing it with others."

Still, his response to my questions came not that day but two days later. His four-page letter began: "To cut to the chase: (1) My recollections are accurate as to what my father told me about the expedition when I was fourteen years old, in 1968. (2) What he told me was entirely false with regard to those aspects of the story which are of concern to you. . . . The purpose of this response is to account for why my father would make up an elaborate story which simply wasn't true."

Brandon hastened to be sure I understood that his father "was a hard-working pediatrician who no doubt saved hundreds, thousands, of children's lives while he was working in rural India for five years (I was there)." He added that his father had been "a tireless worker, an excellent clinician, and a devoted instructor as a professor of pediatrics. He was cheerful and upbeat in character, and there were few who didn't get along with him."

However, Brandon said, Bill had "a habit of padding his résumé" and of telling his son stories meant to make the father look just the way a son would want his father to look. Brandon explained his father's psychology further: "[Bill] was hypersensitive to any kind of criticism whatsoever, but especially to criticism directed at perceived inadequacies in his performance," just the kind of criticism Neel had likely doled out. "He would brood over any such criticism and wreak vengeance upon the critic, but only in his imagination." This combination of behaviors—of generosity, of résumé padding, of imagining vengeance—covered up "the painful reality . . . that he was a coward."

After putting together various puzzle pieces, Brandon had come to realize that the story of his father Bill standing up to Neel was one among many tall tales his father had told him "for reasons of ego." Brandon recalled in his letter to me that his father had also told him that he knew some Cantonese because he had been in training for intelligence work on the Chinese front during the Second World War. Brandon had since figured out that his father actually knew Chinese because he had had a Chinese girlfriend. "I suppose that when later he was courting my mother, a devoutly conservative Christian, he needed to have a logical explanation as to why he knew Cantonese, an explanation that did not involve him living in sin with a Chinese girlfriend." Brandon, ever the student of literature, suggested, "Perhaps the simplest, most direct way to an understanding of my father is to be found in James Thurber's short story, 'The Secret Life of Walter Mitty,' a favorite story of my father's."

Brandon was sure his father never meant his story of standing up to an amoral Neel to hurt anybody. Brandon suspected the impulse for the story came from Bill's perceiving a slight from Neel—probably when Neel would have had to "put him in his place: that he was not Neel's colleague, that Neel was his boss. . . . Regardless of how tactfully Neel may have phrased it (or not), my father's inevitable emotional response would have been a sense of deep humiliation accompanied by anger and resentment." Brandon continued:

> After he returned home, he told me a lengthy, detailed fiction in which Neel had proved to be a villain and he himself was the hero of the occasion. . . . My father was a hero, and an unsung hero at that! At fourteen years of age, I was sufficiently callow to believe it all. I am certain that I am the only one to whom he ever told this fiction. Telling it once was sufficient to vent most of his continuing sense of mortification and anger. I suspect he felt confident that nothing would ever come of his telling me this outrageously false account, and he was very nearly right.

If Turner and Tierney had not moved to develop a story featuring Neel as a Nazi-like eugenicist, Bill Centerwall's story would have rested quietly in the memory of his son. Brandon would never have formally supplied his father's story to Neel's critics. Brandon might never have realized that the story was but one instance of his father telling tales. But now Brandon had to conclude in his letter to me, "So the story was false in all pertinent aspects. It makes me heartsick to have to write such things about my father. I will stop now."

Now, a few months later, sitting in Frechione's Pittsburgh office, I had to tell Frechione—the man who had tried to help Tierney by recording on tape Brandon Centerwall's memory—that Brandon had withdrawn the story. As I recall, I did not go into the detail of Brandon's letter, but simply let Frechione know that Brandon had retracted his belief in his father's story about the epidemic and that I really didn't think there was anything left in this matter as a strike against Neel.

When I'd made my plans to go to Pittsburgh, I'd taken with me Patrick Tierney's parents' address, the address where he seemed to be living. Tierney had not been answering my requests for interviews. I hadn't been sure when I'd entered Pittsburgh whether I should go to the Tierney house and knock on the door to see if maybe he hadn't gotten my requests. Finally, in talking with Frechione, it became clear that Tierney knew I was trying to reach him and didn't want to talk to me.

However, before I left Pittsburgh, Frechione put a bug in my ear about something that gave me a roaring headache on the drive back to East Lansing. He said that, since *Darkness in El Dorado*, Tierney had been teaming up with Andrew Wakefield—the discredited physician who claimed the measles, mumps, and rubella (MMR) vaccine could cause autism—apparently on new work about the supposed dangers of contemporary vaccines. *What if*, I thought to myself as I drove west on I-80 across Ohio, *what if people take Tierney as seriously this time around as they did the last time? How many people will he hurt?*

After several consultations with advisors about my ethical obligations, I contacted the administration of the University of Pittsburgh to ask why they continued to allow a university affiliation for a man whose work had been shown by numerous major scholarly associations to be fundamentally inaccurate. What about their responsibility to others, including to the public and to Pitt scholars with good names? After some back-and-forth, I received a message indicating that Patrick

Tierney no longer had an active affiliation with the University of Pittsburgh.

IN THE SUMMER of 2009, when I took a short break from this project to visit my family on Long Island, I decided to take the train from New York down to Philadelphia for a day to spend a few hours visiting the archives of the American Philosophical Society to look at Neel's papers. In advance, I let the APS archivist now in charge of the papers, Charles Greifenstein, know that I'd be coming, as you do when you're making a visit like this. I told him that on this visit I just wanted to get a feel for the papers, to know what kinds of sources Susan Lindee and APS archivist Robert Cox had worked with as they had responded to Tierney's claims about James Neel. When I arrived, Charlie took me down to the stacks, and explained that, given the volume of paper Neel bequeathed to the APS, his materials had not yet been completely cataloged. Nevertheless, Charlie had a good sense of what was where. I asked him to give me a box or two from the 1960s, around the time of the ill-fated expedition.

We took two boxes back upstairs to the reading room, where like all other prescreened visitors, I'd be closely watched over by multiple cameras and security personnel charged with making sure patrons don't steal, damage, or insert anything. I started to peruse the papers. One box turned out to contain some marvelous correspondence between Napoleon Chagnon and James Neel, and also between Carlene Chagnon and Neel's secretary. During Nap's early fieldwork, Carlene had gone down to Venezuela with him, bringing their young son and daughter along. Carlene ended up being mostly holed up with their babies in a kind of safe house for Americans while Nap trudged off to the field for long periods of time. Carlene had on occasion taken over correspondence with the University of Michigan, keeping track of grant deadlines and such. As I read these onionskin letters, in which Nap and Carlene wrote to the very edges of the page, presumably to save precious postage, I remembered the two of them telling me about that trip while we were standing around in their kitchen. Nap still felt guilty, nearly fifty years later, for bringing Carlene and the children down in what turned out to be seriously trying circumstances.

I found particularly telling the correspondence between Chagnon and Neel, who was for some of this time off in Japan doing post–atomic bomb research. In the letters, Dr. Neel was trying to give the young anthropologist advice about how to manage in the field, but it was obvious that Chagnon was trying to indicate, as respectfully as possible, that he really didn't need Neel's advice. Chagnon had become so quickly and so thoroughly immersed in field life among the Yanomamö that he already knew far more about them than Neel.

I also found in one of the boxes a letter from James Neel to Mr. Hobert E. Lowrance of the Missionary Aviation Fellowship. The MAF—a team of pilots with small planes—was sometimes hired by Neel and others to get people in and out of the field. The date of the letter caught my attention: April 4, 1968. This letter would have been written not long after Neel had returned from the epidemic.

"Thank you for the letter of March the 25th," Neel began. "We too regret that you were not yet again operational in Venezuela during much of our field work. All of our flying which was not courtesy of the Venezuelan Air Force was done with Mr. Boris Kaminski, whom we found to be both reliable and gentlemanly, although, of course, somewhat more costly than if we had been able to fly with MAF." Neel went on:

> As you may have heard, this particular period of field work was full of unscheduled events. We know from our previous studies how susceptible the Yanomamö were to measles, and had brought with us measles vaccine, which we had thought to administer towards the end of our scientific studies. However, our arrival in the field coincided with the introduction of the disease itself, and we found ourselves in something of a

> race to get the vaccine in ahead of contact with the real disease.

Neel ended the one-page letter by working on arrangements for future trips. He told Lowrance: "Since one of our objectives is to obtain certain biological specimens from the Indians as soon after contact as possible, I hope to get two men in there next year. I would hope at the same time that we will again be in a position to vaccinate for measles, as a small contribution to making the transition of the Indian from the Stone Age to the Atomic Age a little bit smoother."

Not only had Neel not purposefully harmed the Yanomamö; he had been actively working to protect them from what he knew would inevitably come. This accorded with what Lindee and others had shown. I called Charlie over and showed him this letter, so nice a summary proof of the reality did it make. He was visibly excited by it, and immediately took it to make me a copy on bright blue paper. (High-security archives make copies on colored paper so you can't "accidentally" walk off with an original.) But while he went off to make the copy, I found myself in pain.

Here, within an hour of arriving at the APS archive, I had come upon a letter clearly showing that Neel had been bringing in vaccines because he was concerned about the Yanomamö's vulnerability to measles. The letter seemed to show the epidemic had started just before he got there (something I knew the missionary-anthropologist Thomas Headland had confirmed by independent documentation). It showed that he had done what he could to get ahead of it. It showed he was making plans to get more Yanomamö vaccinated, in order to protect them. *I had found this additional piece of clear evidence in an hour.* Why the hell hadn't the *New Yorker*'s fact-checkers? And how could this sort of letter—on top of what Lindee, Cox, Headland, and all the rest had found—have been so utterly ignored by the AAA as they plowed ahead for *years* in an attempted prosecution of Neel based on Tierney's worthless indictment?

In my interviews with people who had participated in the AAA's *Darkness* doings, I hoped to understand this, to get some explanation about why the AAA had so radically differed in its approach from all the other scientific societies that had weighed in. The only explanation I could come up with was that (a) Terence Turner pushed and got others to push; (b) no one wanted to "censor" persons who claimed to be speaking on behalf of vulnerable indigenous peoples, and (c) Chagnon was rough around the edges. In other words, an old white guy known for drinking beer and swearing who had made his name relating stories about vulnerable native peoples getting stoned, getting laid, and getting killed—well, such a guy was understood to have forsaken all his rights to a fair trial at a place like the ultraliberal AAA.

This whole scene was just so disturbing. Everything I found indicated that the leaders of the AAA had to have known early on that Tierney's work was riddled with errors. Yet they had proceeded anyway. To justify it all, they had first formed the Peacock Commission—a panel named after the chair, James Peacock, a former president of the AAA—to come up with a plan for responding to Tierney's allegations. The Peacock Commission issued what amounted to a sealed indictment; no one outside the AAA's upper echelons was even allowed to see exactly what the Peacock Commission was telling the AAA leadership. AAA rank and file were allowed to know only that the Peacock Report had been used as the justification by the AAA leadership to form the El Dorado Task Force. Naturally I wanted to see this secret Peacock Report, and I was glad when I was given a copy of it by Chagnon's collaborator, Raymond Hames, a task force member who had been added late under pressure from Chagnon's supporters and had later resigned. (When Hames resigned, he claimed it was because of a conflict of interest, but he admitted to me in our interview that it was actually because he saw the task force becoming a train wreck where facts were concerned.) When Hames e-mailed me a PDF of his copy of the Peacock Report, I was stunned. It just summarized what Tierney's book said, and it contained no logical argument for why the AAA should proceed. They had so rushed it that not all members even had had time to approve the final version before it was submitted with their names. *This* was the justification for the AAA's El Dorado Task Force and the surrounding free-for-all?

Yet it got worse. As I dug around, interviewing the members of the task force and uncovering internal documentation, it became obvious that several of the people leading the AAA's "inquiry" *absolutely knew* that Tierney's book was a house of cards. Some did try to do the right thing. The Wisconsin biological anthropologist Trudy Turner (no relation to Terence Turner), who served on the task force, desperately tried to use her position from within to assert the documented truth about the measles outbreak and the vaccines. But she was up against others who simply didn't want to deal with the facts. When I interviewed her, Trudy Turner told me that Ray Hames's resignation "left me in a very difficult position. If I had bailed, who would have been left [to defend the truth]? I felt like I couldn't. Then they could have written anything they wanted about Neel."

Even knowing that this whole show had been run like a junior-high moot court with no adult supervision, I could hardly believe it when, in our phone interview, Peacock Commission and task force member Janet Chernela (whom Chagnon wryly nicknamed "Chernobyl") told me, "Nobody took Tierney's book's claims seriously. I was surprised that James Peacock, who is a very careful and fair person, favored going forward with the task force." Yet go forth it did, *with Chernela's cooperation on two commissions for two years.* Indeed, Chernela had actively participated in facilitating "testimony" from a Yanomamö spokesperson who claimed that Chagnon had offered to pay his subjects to kill each other and had offered to pay per killing. Chernela allowed reproduction of this utterly baseless claim in the Task Force Report, and thus also over the Internet.

Meanwhile, my interview with the chair of the task force, Jane Hill, made my head positively spin. Hill seemed to want simultaneously to give me the conclusions of an ethics investigation—saying, "The final report did include an evaluation of Chagnon. We dismissed the charges against Neel"—while claiming to me that "we did not make an ethical accusation against [Chagnon]. But not everyone read it that way." She added, "I think he could have gotten a lot worse from us than he did." Worse?! Hill told me that the whole thing had been hard on her, and I had the sense she really didn't want to revisit it. But I felt she never really understood how hard it had been on the Chagnon and Neel families. When I asked her why Chagnon was never formally invited to defend himself, she answered, "A decision was made not to talk to him. I don't remember the circumstances."

Everything became a bit clearer when I received a batch of photocopies from Sarah Hrdy, a primatologist who had been invited to join the task force but had declined. In the stack of documents Hrdy sent me, I found what was supposed to have been a confidential e-mail exchange dated April 2002 between Hrdy and Jane Hill, the chair of the task force. (To Hill's credit, when I asked, she gave me permission to quote her message in my work.) A good friend of Hames and a supporter of Chagnon, Hrdy had written to Hill, after Hames resigned, to express concern over what was going on. Hill replied:

> Burn this message. The book is just a piece of sleaze, that's all there is to it (some cosmetic language will be used in the report, but we *all* agree on that). But I think the AAA had to do something because I really think that the future of work by anthropologists with indigenous peoples in Latin America—with a high potential to do good—was put seriously at risk by its accusations, and silence on the part of the AAA would have been interpreted as either assent or cowardice. Whether we're doing the right thing will have to be judged by posterity.

So this was the heart of it: The task force knew "the book is just a piece of sleaze," but in their view, Chagnon had to be strung up to save anthropology from Tierney's bad rap.

Meanwhile, another source gave me a tip that Louise Lamphere, then the president of the AAA, had actually suggested that Chagnon's own university investigate or censure him. This was pretty extraordinary—the head of a scholarly association reaching out to push a university to investigate or censure a scholar even as other scientific societies were denouncing the accuser in chief. I wrote to Francesca Bray, who then was chair of Anthropology at UCSB, Chagnon's academic home.

One remark from her pretty much said it all: "I never thought I'd feel sorry for Napoleon Chagnon." Bray recalled to me that Lamphere had called her at home on a Sunday morning to suggest that UCSB might want to do something about Chagnon. Trusting her perceptions, I pushed her to give me a sense of how she saw Chagnon, after years of dealing with him personally. In a follow-up e-mail to me, she wrote this: "He certainly could be rough, but as a colleague at UCSB he was (if often provocative) reliable, straightforward, funny, and generous in his support to colleagues even when he disagreed with their theoretical bent."

That matched my sense. And, boy, I never thought I'd feel sorry for someone like Napoleon Chagnon either.

THERE SEEMED ONE most logical place to present my research on the Tierney-Chagnon controversy: the annual meeting of the American Anthropological Association. Because of the way academic conference cycles work, I had to propose to present my work at the December 2009 meeting—scheduled for Philadelphia, of all places—many months before I knew what I'd ultimately find. So sometime early in 2009, I wrote a relatively vague proposal that promised simply to report in December on my findings. As part of that proposal, I invited Terence Turner and Tom Gregor to provide formal responses. (This was Gregor of the Gregor and Gross duo behind the referendum rescinding acceptance of the Task Force Report.) I didn't want to see Turner shut out of the panel as I had been at the National Women's Studies Association panel about my work on the Bailey history. And for all I knew as I wrote my proposal in the spring, by the time of the December meeting, I might find evidence of serious ethics violations by Chagnon and Neel, which was why I wanted to be sure either Gregor or Gross could also, if necessary, answer for what they had missed. I promised Turner and Gregor that I would get my presentation paper to them at least ten days in advance, so that they could fully prepare their responses.

When the time came, in November 2009, to write it all up, I knew I also had to show what I had to Nap and Carlene, in part so that we could discuss how much they were willing to let me say about what had happened to their family because of all this. I realized that, during what might be a stressful visit, Nap and Carlene might both benefit from having a good friend along, someone who might help them remember the better times. So I asked Raymond Hames (who had been friends with them for over thirty years) to fly up from Nebraska and join us. I also figured that if I got into an argument with Chagnon—we did have a tendency to snap at each other over various points and issues of methodology—Hames could help me win. I'd pick up Hames at the Grand Rapids airport on the way north and drop him back at the airport on my return home.

I'd met Hames in person a few months earlier, at the Human Behavior and Evolution Society meeting in Fullerton, California. HBES is the home organization of sociobiologists, or more specifically evolutionary anthropologists and evolutionary psychologists. Chagnon had been one of the founders. Never in my life would I have thought I'd want to go to an HBES meeting, much less present at one. But after my first set of interviews with Chagnon, it occurred to me that he and I could both benefit from an HBES session about my work. I could run my initial ideas past the scientists most directly implicated by this project and try to encourage them to be more careful and more organized in defense, and Chagnon could come out of his Michigan exile and be back in Southern California with his old buddies.

Given that I still had a lot of work to do on the *Darkness* history in May, and given that I'd have only about a half hour to present at HBES, I proposed to relate there only the most basic outline of each controversy and then to point to the common themes in these matters. We would then have several scientists respond, including Chagnon, Mike Bailey, Randy Thornhill, and Craig Palmer. Chagnon and I coordinated our flights so that we would both leave out of Grand Rapids. From the start of the trip, his excitement over seeing numerous old friends made him giddy, and as a result, he was driving me nuts. First off, he kept joking to everyone that I was his assistant. When he told this to the TSA agent at Grand Rapids, I explained that I was actually a full professor at Northwestern's medical school.

"I get very qualified assistants," Chagnon loudly whispered to the TSA guy.

Then on the plane, Chagnon proceeded to tell me what an overly sensitive person might construe as a racist joke. It was kind of funny, but I told him to keep it down, at which point he got into a joking argument with me about my racist assumption that the joke was racist. Later, while talking with me about his controversy, he said rather audibly, "You know, if you think about it, there was a Jewish conspiracy at work in my case!"

"What the hell are you talking about?" I asked him, turning red.

"Think about it!" he answered. "Hames, Gregor, and Gross—practically everybody who helped me is Jewish!"

When it came time to give my talk to the scientists assembled at HBES, I tried to summarize as best I could what the themes were in these controversies. I talked about the problems of personalities. I talked about identity politics being confused with truth. I talked about how working on sex makes you more likely to get in trouble than just about any other topic. I talked about how Galilean types—men and women who are smart, egotistical, innovative, and know they're right—tend to get in trouble, especially when there's a narcissistic nemesis around. I talked about the way Galilean types tend to believe that the truth will save them, and to *insist* on the truth even when giving up on it might reduce their suffering. I suggested that scientists be *offensive* in their techniques instead of just defensive; specifically I suggested they watch their language—not use trouble-attracting phrases like "good genes" as a stand-in for "genes that make one more reproductively fit." I suggested they try to engage the audiences implicated by their work early in the process. I suggested they support each other when baseless charges were thrown about, and not assume that just because a colleague is engulfed in smoke, that he or she has actually set a fire.

Throughout the HBES meeting, both in my session and outside it, various scientists tried to tell me what they thought Chagnon should have done to protect himself. He should have sued. He should have had UCSB (where he was emeritus) defend him the way Michigan had defended him and Neel. He should have gone to the AAA and yelled at them in person. All these suggestions reminded me of the reaction people naturally have when someone gets cancer. We try to find a way to blame the individual, I suppose in a psychological attempt to tell ourselves it couldn't happen to us—*we* would be more careful; *we* would not find ourselves in such a situation.

The truth was that there wasn't a lot Chagnon could have done, because no one had been through anything this crazy before. Everybody had just improvised. At UCSB, three of Chagnon's colleagues had mounted an exhaustive defense, and at the University of Michigan, the home base for Neel and Chagnon during the epidemic, the provost's office had moved extraordinarily to issue a formal defense against Tierney. That defense had been largely organized and written by Ed Goldman, a lawyer at the university. Goldman explained to me that the University of Michigan had reviewed the facts in the matter and had announced the obvious conclusion: Tierney's book was wrong—and insulting. Joel Howell, a university physician and historian of medicine (and friend of mine), had helped Goldman draft the rebuttal to Tierney, the historian Susan Lindee had helped, and Goldman—with the brilliance of a defense attorney—had taken the lead in writing the plain-language point-by-point defense that would emerge in the provost's name. The University of Michigan had done what it could to set the record straight.

I knew that something just as extraordinary had happened for Chuck Roselli, a neuroscience researcher at Oregon Health & Science University (OHSU), just a year or so back, when People for the Ethical Treatment of Animals (PETA) had decided to go after Roselli for his studies of male sheep that prefer sex with other males. PETA got the gay and lesbian activist world all up in arms, claiming that Roselli's work was intended to become the basis for a kind of eugenics that would prevent homosexuality in the human population. Even Martina Navratilova ended up going after Roselli in the press. Roselli's university president received something like *twenty thousand e-mails* objecting to Roselli's work and calling for his firing.

Luckily, OHSU had a public relations guy up to the challenge, a man named Jim Newman. Newman went into action, setting up an auto-response for the e-mails, answering with the facts about Roselli's scientific work. If people replied to the auto-response, Newman answered, engaging them, to try to get them to understand the reality. Roselli himself did some media, but as he quickly started burning out from the craziness, Newman took over to some extent, giving extensive interviews and constantly bringing people back to the facts. Newman engaged one blogger after another, writing comments in his real name, pointing people to a page he had set up at OHSU aimed at challenging misrepresentations.

For months, Newman worked on this and virtually nothing else around the clock. Although the university supported his efforts, Newman told me, "For me, it was more about defending Chuck and Chuck's research more than the university. We wanted to respond because of what they were saying about this researcher, and we have to defend researchers this way."

You know what? Newman turned the damned thing around. He even managed to get some gay and lesbian activists to say they were angry at PETA for misleading them! When I interviewed Newman and asked him how he thought about the whole thing, it was clear that for him it was a matter of supporting a researcher who was just trying to do good science: "My concern, really, was that Chuck was spending a lot of his time doing this [dealing with the controversy] instead of his work. But at one point he was able to kind of say, 'OK, I think I can get back to work.'" And that, to Newman, was a sign that they had won—because scientific research could now resume.

But how many Jim Newmans were there out there? One, so far as I knew. One Jim Newman, one Ed Goldman. Well, also one Trudy Turner, one Susan Lindee, one Tom Gregor, one Dan Gross. It was true that Chagnon had not had the benefit of all the things that might have helped. It was true the AAA had gone utterly out of control. But it was also true there were examples of good people who had resisted and fought well for true scholarship in many of these messes. I tried, in the little time I had, to impress this upon the scientists at HBES. I said they had to stand up, for themselves, but especially for each other and for the facts. They had to ask to have Ed Goldmans and Jim Newmans available, and they had to call on the Susan Lindees and Alice Dregers as necessary to see that science was not abused in the service of identity politics (and personal vendettas) run wild.

After I finished my prepared spiel, Craig Palmer and Randy Thornhill each talked a little about what they thought mattered. Mike Bailey got up and said that he agreed with me that scientists should be offensive—and he clarified that to mean that offending people was a sign of doing important work. (I groaned; exactly *not* what I meant, but there was that Galilean personality.) Chagnon then got up. He started by talking a little more about what he had been through. And as he did, as seemed inevitable, he started to do that choking up thing again. Which of course choked me up, yet again.

"Damn it," he said to me in the elevator afterward, "every time I get near you, Dreger, I start bawling like a girl!"

"But, Dr. Chagnon," I answered, mock-sniffling, "that's my claim to fame."

Indeed, his emotion had been noticeable. Several scientists came up to me at the banquet that evening and told me how intense it had been to see the ultimate alpha male choke up like that. "I didn't really understand what he'd been through," one said to me. "I guess I still don't understand it, but it must have been really bad."

By the time months later, when I went up to Traverse City, this time with Ray Hames to tell Nap and Carlene what I planned to say at the AAA, I knew exactly how bad it had been. But this time there were no tears. We did a fairly clinical run-through of my work. Nap caught one date error, the only item I had not had time to fact-check before leaving home. And we talked about what the AAA session might be like.

At one point in the visit, I had to dip out of our group conversations to take a call. I sheepishly explained to Nap that a fact-checker from the *New Yorker* needed to verify some material. I'd been helping a writer there with a piece on intersex.

"By all means," Nap said to me, with understandable sarcasm in his voice. "Imagine if they had bothered to fact-check Tierney's story on me."

ON THE PLANE to Philadelphia to present my findings at the AAA in December 2009, it suddenly became clear I was coming down with a serious respiratory virus. As we taxied from runway to gate, I kept doing the math hoping that if it was the H1N1 flu (which was going around that year), there was no way I could have exposed Nap to it a few weeks before. With the significant respiratory problems he had developed from years spent sitting around cooking fires and smoking unfiltered Pall Malls, the flu might do him in.

By the time of my AAA session the next day, I had started to lose my voice, and I felt increasingly feverish. The adrenaline that came with the session helped a bit. Although I almost never read a paper for a presentation, here I had no choice but to read, to make sure that the work I presented was exactly what I had provided to Turner and Gregor ten days in advance, so that there was no chance they would feel ambushed. Because of what I'd found, the AAA audience had no trouble staying interested. When I finished, Gregor and Turner each got the fifteen minutes at the podium I'd arranged. (Turner spent much of his allotted time complaining that I'd given him only fifteen minutes.) Then the moderator, a well-respected anthropologist peacemaker whom I'd chosen knowing he'd be fair, did exactly what I'd asked of him: During the discussion period, he called first on anyone I had criticized who now wanted to speak. Although by the end I wanted nothing more than to go to bed, I stuck around to talk to the reporters from *Science*, the *Chronicle of Higher Ed*, and *Inside Higher Ed*, who were looking to get copies of my paper.

By the time I got home the next day, I had a full-blown fever and a terrible cough, a cough that got worse and worse over the next several days. (It turned out to be pertussis. My doctor had forgotten to vaccinate me.) As I fell in and out of sleep, I kept thinking back to the one hostile question I'd gotten at HBES, from a man who accused Bailey of being offensive in his work. He had asked me, near the end of our session, whether I thought that perhaps the Galilean personality I was describing also applied to me. I had answered promptly: yes.

And now I knew, better than ever. *Yes*. Pugnacious, articulate, politically incorrect, and firmly centered in the belief that truth will save me, will have to save us all. Right in the fight but never infallible. Yes, I thought: In Chagnon, I have met the ghost of Galileo. And he is me. He must be all of us.

CHAPTER 7

RISKY BUSINESS

As I LAY about recovering from whooping cough in late 2009, reflecting on all I had learned in my journeys, I had a vision of how much *easier* social justice work around scientific research might become if it *were* consistently evidence-based. Scientists, the vast majority of whom I now understood to care deeply about social justice, would have to respect evidence-based activism. Maybe if everybody just agreed to discuss what we really knew—rather than imagining, assuming, and suspecting based on loyalties to particular theories or persons—disputes could be sorted out peaceably. Researchers and advocates could come together, look at the facts, and—as in the case of climate change, AIDS research, and intersex care—the great majority of researchers and activists would agree on what had happened and what needed to happen. The light of many minds would show a way forward. After carefully investigating and getting reasonable people to see what had actually happened in the Bailey and Chagnon cases, I felt downright optimistic that if we all simply agreed to talk through what we knew and didn't know, *true* cases of injustice would be relatively easy to spot, expose, stop, and ultimately prevent.

As if by providence, at that moment—just as I was feeling tremendous clarity about what activism should and shouldn't look like—a small group of my old intersex-rights allies, including a couple of clinicians with whom I'd remained close, called me to ask me to help with an ongoing travesty of justice. The crux of the matter was this: A major clinician-researcher in the intersex field (someone we all knew well) had been promoting a high-risk drug regimen to pregnant women who, along with their mates, were genetic carriers for a particular intersex condition, one caused by elevated levels of androgens in female children. The idea behind the intervention was this: If, through genetic screening, a clinician identified a woman as being at risk of having a daughter with this condition, then as soon as the woman got pregnant, a doctor would start dosing her with the steroid dexamethasone. If all went as planned, the dexamethasone would dampen the effect of the androgens, preventing intersex development in the daughters.

The major proponent of this fetal intervention, Dr. Maria New—a distinguished member of the National Academy of Sciences and a charismatic pediatric endocrinologist still actively working in her eighties—was going directly to the support groups for families with this condition to strongly recommend the intervention to them. She was selling parents on this "treatment" by claiming that "with nearly 20 years' experience, the treatment has been found safe for mother and child." This wording made it sound as though the Food and Drug Administration (FDA) had approved this use, when the FDA hadn't. While doctors are allowed to prescribe drugs "off-label," i.e., for uses the FDA has not approved, doing so means acting without the benefit of extensive FDA review of a use. In fact, the medical-scientific literature was utterly devoid of well-controlled studies of efficacy and long-term safety of prenatal dexamethasone for intersex prevention. Doctors didn't know *what* the intervention really did in the long term, although many were very worried; dexamethasone was being used specifically because it had the power to cross the placenta and permanently alter the course of fetal development. Indeed, lab-based studies of prenatal dexamethasone exposure in animals, including nonhuman primates, suggested that dexamethasone exposure could shift fetal development in deeply unpredictable ways, including by changing brain development.

It got worse. Even while Dr. New was describing this use as safe and effective, openly taking credit for having gotten over six hundred pregnant women to follow this prenatal regimen, and boasting that she had at Mount Sinai School of Medicine in New York "the only clinic in the United States which routinely provides this service," she was *at the same time* taking National Institutes of Health (NIH) funding to study, retrospectively, what had really happened to children and mothers who had been exposed. To get the NIH funding, Dr. New told the agency something very different from what she said in her public promotions of prenatal dex: "The long-term effects of dexamethasone on the children who received it as fetuses and on mothers who were exposed to it while they were pregnant *have not been determined.*" She told the NIH that her large "accumulated" clinical population of exposed women and children made her especially well positioned to do this research.

Telling pregnant women a fetus-altering drug is "safe and effective" and then using them to get grants to see if it really *is* safe or effective? The failure of ethics in this case seemed so appalling and so obvious to me that it seemed a perfect case study of how to spot and stop a true and significant case of abuse of research subjects. And it involved the very issue around which most of my work had revolved: the often-harmful, non-evidence-based medical quest to "normalize" children. Here again was a situation where doctors, motivated by not-unreasonable fears of social stigma, were driving forth an unscientific and even unethical system of treatment—in this case, risking the health of pregnant women and their babies, completely outside of modern scientific standards for pharmaceuticals. Indeed, after I did a little more reading and investigating, the situation seemed so clear that I thought we could and should use this as a demonstration of evidence-based social justice protest—a model of how you appropriately use evidence, the press, and the Internet to reign in abusive research. This case could serve to show scientists why activists were very reasonable to still be vigilant about ethics violations—because terrible abuse really was happening to some vulnerable subjects—and to show activists why and how we should work to push for better evidence. Moreover, because the researchers calling for my help were calling me in *as* an activist, I thought this would be a great example of collaborative work between the two camps.

Yes, it would be relatively easy! The source of our outrage, the reason for our collaboration across the researcher-activist divide—it would all be obvious to everyone. The press, the universities, and the government would do what they're supposed to do. People weighing in on the Internet would behave well. Everyone would see that *this* is how you do it: You push together for the truth.

Looking back, I'm guessing I was high on cough medicine. That said, even if I hadn't been, I would have taken on this cause. It was too important to say no. This was one quest for pediatric "normalization" run completely amok. And although this work—which I guessed at the outset would take only three months—ended up consuming the next three years of my life, nearly crushing my reputation and my spirit, it did end up teaching me the last things I needed to learn from my long journey: how badly most people want simple stories of male and female, nature and nurture, good and evil; how the Internet has gutted the Fourth Estate; how the government is made up of fallible and occasionally disappointing humans; and why, more than ever before, democracies must aggressively protect *good* research.

AT THIS POINT in my intellectual development, after witnessing the personal consequences to individuals attacked for controversial research, to say that I was reluctant to call out a single researcher would have constituted a substantial understatement. But after my colleagues brought this case to my attention, I poked around and could quickly see why they were focusing specifically on Maria New. Dr. New, I realized uncomfortably, had made herself the eight-hundred-pound gorilla of prenatal dexamethasone for intersex prevention. She really stood alone. There had been a number of major pediatric researchers, including prominent physicians with whom I had for years argued over intersex care, trying to put a stop to her misleading promotion of this intervention as having been found safe and effective. A quick search confirmed why they were a bit frantic: There remained a complete absence of any properly controlled scientific studies demonstrating efficacy or safety in humans. There hadn't even been any animal modeling of this use published. The animal studies that *had* looked at prenatal dexamethasone were designed simply to find adverse effects of prenatal exposure to dexamethasone in general; no animal studies had been published specifically looking at whether you really could stop intersex development via prenatal dosing with steroids. The number-one rule of pharmaceutical research? Before you touch a human, see if the intervention works in animals. This was *especially* true for fetal experimentation. Yet such animal modeling hadn't been done, and New's retrospective sampling methodology suggested that the results of her NIH-funded study would be of low scientific quality. While what she was doing didn't appear to be illegal according to FDA prescribing rules, *ethically* it was extremely problematic.

Meanwhile, alarming data were emerging from a Swedish research team that had been tracking children there carefully from the start of prenatal dexamethasone exposure straight through later childhood. The Swedish data suggested that prenatally exposed children were showing increased rates of problems with memory and with social anxiety—signs that the intervention might have adverse brain effects. And the rest of their bodies? The Swedish group had found cases of growth failure and of delayed psychomotor development among the children prenatally exposed, suggesting unintended effects on the rest of the body as well.

Dr. New had not been studying the effects of exposure nearly as carefully as the Swedish team, even though she had taken credit for exposing well over *fifteen times* as many mothers and children. I did the math to figure how many pregnant women per month might be getting sucked in to this unethical cycle of ill-informed "treatment" and federal follow-up research grants, knowing the frequency of the condition at issue (somewhere between 1 in 10,000 and 1 in 15,000 births) and knowing that New's clinic drew patients from around the world. It could be a dozen per month. Then I mentally put myself in the position of these pregnant women, most of whom would know only what was told to them by the intervention's reassuring promoter. Then I thought about what New was doing—telling the families the intervention had been found safe while using previously exposed families to get NIH funding to see if it was safe.

Then I went back to my friends' central question: What should we do?

You'd think that if anybody was qualified to answer the question *How do you stop a scientific researcher?*, it would've been me at that moment. The problem was that my work had only uncovered all the *wrong* ways to do it. What the hell was the *right* way?

TO UNDERSTAND HOW my colleagues and I were processing this scene, one needs first to know a little biochemistry, a little developmental biology, and a little history.

In spite of what popular culture might lead you to believe, males and females usually make the same types of hormones. In addition to making the hormones usually associated with their sexes, male bodies also make the "female hormone" estrogen and female bodies also make the "male hormone" testosterone. What determines male and female biology is, in part, the different *levels* of sex hormones that usually occur in males and females during sex development. Being estrogen-heavy will usually make you more female-typical, while being testosterone-heavy will usually make you more male-typical.

Ovaries equal more estrogen, and testes equal more testosterone. That's why most females are feminine and most males are masculine. But again contrary to pop culture's representations of sex, the ovaries and testes aren't the only organs in the human body that produce sex hormones. Some sex hormones are also made by our adrenals, a pair of glands that sit just above our kidneys. These glands make androgens, a group of hormones sometimes called masculinizing hormones: androgens can push the development of sex differentiated-tissues (like genital tissue, breast tissue, and some brain cells) in a more male-typical direction during sex development, even if you have ovaries. Perhaps you already see where this is going: There exists a condition called congenital adrenal hyperplasia (CAH) that results in higher-than-normal levels of androgens. If CAH occurs in a genetically *female* fetus, in spite of the fact that the fetus has XX chromosomes, ovaries, and a uterus, the higher-than-normal level of androgens can lead that fetus to be born masculinized in terms of genital appearance. CAH is, in fact, the most common cause of congenital ambiguous genitalia in genetic females.

The reason it's impossible to know the efficacy of prenatal dexamethasone for CAH without a blinded, controlled drug treatment trial—something that has never been done—is because there is actually substantial natural variation in genital formation of CAH-affected females, ranging all the way from nearly typical female to nearly typical male. The natural variation among XX children with CAH means that a child with the condition may be born with a typical clitoris, a large clitoris, or in very extreme cases even one that has the urethra running part or all the way through it, just like a penis. The labia majora may be separated in the typical female fashion or may be joined at birth, looking much like a male scrotum. The XX child with CAH may have a normal vagina, may have a vagina that will not allow normal menstruation and vaginal intercourse, or in very rare cases may be essentially missing a vagina altogether. Sometimes a CAH-affected XX child will have something called a urogenital sinus, wherein the urethra and vagina develop in such a way that they meet, creating the potential for urine and menses backwash and subsequent infections.

Like genitals, brains are also shaped by prenatal androgens, so it should not surprise us that many CAH-affected girls' and women's behaviors and interests show what researchers call *behavioral* masculinization. Although there is a lot of variation in this population, CAH-affected girls are more likely than average girls to be interested in boy-typical toys, like cars and trucks, than in girl-typical toys, like baby dolls. They're more likely to want to play with boys and are more interested in rough-and-tumble play. CAH-affected women are more likely to be bisexual or lesbian, and less likely to be interested in becoming mothers. A sizable percentage of genetic females with CAH—as much as 5 percent—ultimately identify as male in terms of their gender identities. The behavioral outcomes in this population strongly suggest that gendered behaviors and sexual orientation have a biological inborn component, and in CAH, females are skewed in a male-typical direction.

CAH can actually show up in both males and females, but only in the females can it cause intersex development. (Extra androgens in a developing male won't make his sex ambiguous.) Notably, in both males and females, CAH can be a dangerous endocrine disease. Because the adrenal glands are responsible for the body's response to physiological stress, having CAH can mean a poor response to infections and traumatic accidents. In the most extreme forms of CAH, a newborn baby who isn't diagnosed and treated promptly with steroids can go into an adrenal crisis and die within a few days of birth. The potential seriousness of CAH is why all fifty states now employ newborn screening for CAH. If prenatal dexamethasone could prevent or cure CAH—not just prevent intersex development—that would require a recalculation of the ethical equation of using (and pushing) this intervention. Unfortunately, the endocrine disease that is CAH can't be prevented, nor can it be cured. No matter what you do to a CAH-affected fetus prenatally, the medical dangers of CAH will have to be managed postnatally with monitoring and medication. This means that even if prenatal dex prevents intersex development, an exposed child will still face the serious health consequences of CAH.

CAH is a recessive genetic condition, which means that if both parents are carriers for the condition, each of their children will have a one-in-four chance of having CAH. Prenatal screening of a fetus in the second trimester has been used for many years to reveal to parents whether a fetus has CAH. Some couples have historically opted to abort. But in the early 1980s, in France—where, perhaps not coincidentally, by the time you could diagnose a fetus with CAH, an abortion would have been illegal—clinicians figured out that dampening down androgens in utero might prevent prenatal atypical sex development in CAH-affected females. A 1984 paper by these French clinicians published in the *Journal of Pediatrics* showed apparent success in one case using a prenatal course of dexamethasone, and after that documentation of *assumed success in just one human case*, "at-risk" mothers-to-be throughout the world started being offered prenatal dexamethasone.

Right around the time of the French group's publication, Dr. Maria New began making the intervention available at her Cornell University Medical College–based clinic in New York City. (This was years before she landed at Mount Sinai School of Medicine, across town.) Dr. New had become internationally known in medicine and among the CAH-affected population for having improved the health and fertility of countless patients through her research and clinical care. She was the top specialist in CAH care. It was therefore easy for her to reach out and recommend this intervention to large numbers of CAH-affected families and their personal physicians. And she did so with gusto.

But why would anyone take such risks with fetuses? To understand this, you have to understand what a strong revulsion and/or pity some people feel toward a person who is born with a body that falls smack-dab between male and female. Maria New was not alone in feeling that intersex should be prevented if at all possible; many of her colleagues had long had the same general philosophy, knowing as they did that, historically, in most cultures, people have been expected to adhere to one of two sex types, in body and behavior. Although in the younger generation, clinicians have changed their thinking about intersex, homosexuality, and transgender, Maria New, born in 1928, had been working consistently from a very traditional, very conservative approach to sex and gender. Here, for example, is Dr. New speaking to a group of CAH-affected families at a support-group meeting in 2001 recorded on video. The audience sees, on a screen beside Dr. New, the masculinized genitals of a baby girl born with CAH—a large clitoris, fused labia. The genitals look pretty male. Dr. New tells the families:

> The challenge here is . . . to see what could be done to restore this baby to the normal female appearance which would be compatible with her parents presenting her as a girl, with her eventually becoming somebody's wife, and having normal sexual development, and becoming a mother. And [internally] she has all the machinery for motherhood, and therefore nothing should stop that, if we can repair her surgically and help her psychologically to continue to grow and develop as a girl.

The message is clear: Genetic females are supposed to have petite clitorises, grow up to see themselves as typical girls, and become wives and mothers. Since the baby shown in the slide was already born, Dr. New was suggesting to the audience that this particular child would have to have her femininity saved by normalizing genital surgeries. From Dr. New's point of view, better still would be the clinical approach of ensuring a sex-typical girl right from the start of development—with prenatal dexamethasone. Dr. New encouraged the parents at the meeting to see this and to understand how personally motivated she felt to recommend prenatal dexamethasone:

> Now one of the biggest breakthroughs, and I think one of the most satisfying things I've done in my scientific life, was to participate in the program of prenatal diagnosis and

> treatment of congenital adrenal hyperplasia. I didn't start it. It was pioneered by a wonderful scientific investigator in Lyon, France, by the name of Maguelone Forest. She is the one who started the idea that it might be possible to treat the fetus by treating the mother. And what we did in New York is simply to refine it, advance it, and make it more feasible. . . . I'll give you some slides to buttress what I'm saying, [but] the conclusion is, it's effective and it is safe. Notwithstanding any people who criticize me for what I'm doing, I think it's one of the *nicest* things I can do for my patients, which is to prevent ambiguity of the genitalia, so that these children, when they're born, do not have confusion of sex assignment—nobody can call it a boy when it is really a girl, and nobody has to undergo very difficult surgical procedures, the surgery for which is only done well in very few centers.

In making this presentation, Dr. New revealed that she herself is a carrier for CAH, a fact that probably explained how she got into this field of research. The message to the parents would have been clear: I am one of you, and I want only what is best for your children. And good news: "The conclusion is, it's effective and it is safe."

What is dexamethasone? A potent synthetic steroid, dexamethasone is actually used for a wide variety of medical purposes. Doctors ordinarily prescribe it for children and adults to treat inflammation, like the kind you might get with some eye or skin diseases, and to treat various autoimmune diseases, like psoriasis and lupus. Dexamethasone has sometimes also been given in the second or third trimester to pregnant women at risk of giving birth prematurely; that's because some of the steroids known as glucocorticoids—the drug class that includes dexamethasone—can cross the placenta and push a fetus's lungs to mature faster, giving the fetus a better chance at survival if born too soon.

But the prenatal use of dexamethasone at issue in the intersex story aims not at second- or third-trimester fetuses who might otherwise *die*; instead it aims at *first-trimester* female fetuses who might otherwise develop in *sexually atypical* ways. In fact, it'd be more accurate to say the CAH-related use of prenatal dexamethasone is aimed at *embryos*. Humans start to differentiate into male and female types really early, by the seventh week of development. If you want to reengineer human genital development, you have to hit an embryo—through its pregnant mother—with dexamethasone pretty much as soon as you know the mother is pregnant. Usually the parents are identified before pregnancy because they've already had a child born with CAH, but additional "at-risk" parents are picked up because they've decided to undergo preconception genetic screening, and the screening picks up that both parents are carriers. Through the CAH care networks, CAH-affected families steadily learned over the years to start a woman on dexamethasone the moment a pregnancy with CAH risk was confirmed.

The prenatal use of dexamethasone for CAH has always been "off-label," meaning that, while dexamethasone has been approved by the FDA for some specific medical uses (like treatment of inflammation), this has never been one of them. In fact, the FDA has for years classified dexamethasone in pregnancy category C, those drugs for which "animal reproduction studies have shown an adverse effect on the fetus and there are no adequate and well-controlled studies in humans." Why then would doctors still be allowed to use it in pregnant women? Because, as the FDA notes for category C drugs, "potential benefits may warrant use of the drug in pregnant women despite potential risks." Some doctors, like Maria New, decided that the "potential benefit" of intersex prevention justified potential risks.

But one would expect such doctors to be terribly careful with a first-trimester sex-engineering attempt, particularly since the DES disaster. Starting in the late 1930s, physicians sometimes gave pregnant women diethylstilbestrol (DES), a synthetic estrogen, generally to prevent miscarriage. A controlled, blinded study published in 1953 in the *American Journal of Obstetrics and Gynecology* actually showed that DES didn't work for this use; women who were given DES did *not* end up with healthier pregnancies. Nevertheless, obstetricians kept pushing DES on pregnant women, perhaps unaware of the study, perhaps disbelieving it, perhaps thinking DES couldn't hurt. With no good evidence of efficacy or safety, about 5 to 10 million fetuses were exposed to DES by the time this use finally came into serious question in 1971. It was then that clinicians in Boston realized that girls and young women who had been exposed to DES in utero were developing a rare and often fatal vaginal cancer at an alarmingly high rate. They realized the DES link to this strange cancer cluster only because a woman named Penny Stone, a mother of a teenage girl with this rare vaginal cancer, insisted that researchers consider as a possible cause the DES she had taken when pregnant with her daughter Sheila. A few months after the Boston group made its findings known, facing intense pressure from a Congressional investigation, the FDA issued an alert, and the intervention eventually tapered off among obstetricians. Since then, researchers have linked prenatal DES exposure to increased risk of many reproductive cancers, as well as to major fertility problems in prenatally exposed males and females.

The negative effects of DES did not present themselves in obvious ways at birth, which is part of the reason they took so long to be recognized. But as the medical profession learned the hard way, when a drug given during pregnancy is studied unscientifically, even *dramatic* birth defects may not be tied to the drug until thousands of babies are harmed. In the middle of the twentieth century, the widespread use of thalidomide, a sedative and antinausea drug for adults, led to babies exposed in utero being born without limbs or in some cases with flipperlike limbs. It was only in the early 1960s when Frances Oldham Kelsey, a scientist at the FDA, pushed very hard to collect and analyze data that the disaster of thalidomide came to be widely recognized. By that point, about ten thousand children in Europe had already been born with shocking birth defects because their mothers had taken thalidomide when pregnant with them, typically to deal with morning sickness. Yet, until a prominent *Washington Post* article about Kelsey's scientific efforts emerged, the buzz had continued to be that thalidomide was a safe and effective sedative and antinausea medication.

Everybody working in medical research knows that there are lots of formal protection systems in place to prevent another DES or thalidomide. For example, researchers who want to experiment on pregnant women and fetuses are supposed to jump through extraordinary regulatory hoops before being allowed to proceed. There's also supposed to be a kind of *cultural* protection in place from the conventional wisdom developed in response to DES and thalidomide, a conventional wisdom that says you don't mess with fetuses pharmacologically unless you absolutely must, unless, for example, you think the fetus will otherwise die, as in the deployment of steroids for premature birth. Even then, you are supposed to proceed with profound trepidation and enormous scientific care if you know the drug crosses the placental barrier and affects the fetus. DES and thalidomide showed us how many cases of harm it can take, if you're not scientifically careful, to start to learn what effects you're really having when you're exposing fetuses to pharmaceuticals.

So imagine how careful you should be when you're exposing embryos to a drug *with the intention of changing how their tissues will develop.*

And how often does a pharmaceutical intervention have in all cases *only* the effect intended?

Hint: Never.

AS I BEGAN THINKING ABOUT the dex issue, one of the clinicians I was talking to was David Sandberg, a pediatric psychologist from the University of Michigan and Mott Children's Hospital in Ann Arbor, with whom I'd long been collaborating on efforts to push back against risky attempts at medically "normalizing" healthy kids. Neither David nor I saw in theory any deep moral problem with preventing the development of ambiguous genitalia in utero. Neither of us saw intersex genitals as any more special or sacred than any other genitals, and all other things being equal, especially in our still-crazy medical system and in our culture at large, a child would almost certainly be better off born with typical genitals. Doing something to prevent "the need" for a surgeon to come at these children's genitals with a knife? In theory, sounds good.

But as David talked me through what was going on, he made clear that the major concerns he and others had were not with theory but with practice. First off, there was the concern that the drug had to be given so early that it meant women were started on it as early as the fourth or fifth week of pregnancy. At that point, the embryo being targeted would be as small as a sesame seed, and the total dose of glucocorticoids reaching it would probably be sixty to a hundred times the naturally occurring level of glucocorticoids. Much more disturbing, it wasn't at all clear whether the pregnant women understood that doctors had very little idea what this drug might do to them or their children in the long run or that many experts were plenty worried. *Twenty-five years into use, we didn't even have a placebo-controlled study to show it even did what was intended.* All we had were reports from the doctors like Dr. New who had themselves recommended the drug saying they judged the intervention to be fairly effective in reducing genital masculinization in the females. Not terribly objective science, to say the least.

Now data coming out of Sweden were showing statistically significant unintended effects on brain development. There could be lots of other problems we didn't know about yet—and wouldn't know about anytime soon if this experimental intervention was conducted unscientifically and without mothers even realizing that it was experimental. Not knowing they were in what amounted to a big, sloppy experiment, the mothers wouldn't necessarily consider the possibility that their children's physical and cognitive problems might be side effects of prenatal dex. Even if, like Penny Stone, they *did* suspect prenatal drug exposure to be the cause of problems, few would know how formally to report their concerns. In the United States, there was no standardized registry of those exposed in utero, only Dr. New's closed clinical records. Dr. New was supposed to be using her NIH grant to follow up with those treated years earlier, but many families were either missing or choosing not to participate when Dr. New's research team called on them to be in the survey.

Because I teach the history of medicine, my friend David Sandberg didn't have to tell me to think about all the parallels to DES and thalidomide. It seemed to us as if the American clinicians using prenatal dexamethasone for CAH had learned nothing from history. But it was also like they were learning nothing from the contemporary scientific literature. By the time I turned my attention to this, bench researchers studying the physiology of development in nonhuman animals had steadily been reporting that glucocorticoids like dexamethasone, when deployed during fetal development, appeared to lead to "fetal programming" of adult pathologies in various organ systems; in other words, prenatal glucocorticoids could permanently skew the developing body, reprogramming the tissues to develop major diseases like cancer in adulthood. Just like DES.

Listening to David rattle off his concerns, I was reminded that Dr. New had left Cornell's medical school a few years back under the cloud of a major NIH fraud case and had then taken a job in another part of New York City, at Mount Sinai's medical school. In 2005, the *Wall Street Journal* had reported that the NIH grant on which Maria New was the principal investigator—the grant that had included Dr. New's research on prenatal dexamethasone—had been at the center of a fraud investigation that resulted in Cornell paying the NIH a $4.4 million settlement. According to the whistleblower, a Greek pediatric endocrinologist named Kyriakie Sarafoglou, who had been working in Dr. New's unit, there appeared to be "phantom research projects" i.e., research projects that were being funded but were not actually happening. (Perhaps this helped explain why, although New kept promising the NIH and her colleagues extensive follow-up data on long-term outcomes of prenatal dexamethasone exposure, the data were strangely absent in the published literature.) According to Sarafoglou, at first Cornell had tried coming after her for raising an alarm about what she was seeing, even requiring her to submit to a psychiatric exam. Instead Sarafoglou went to the Feds.

As he talked me through what he knew, David also wanted to make sure I understood that it wasn't just the females with CAH who were being exposed. Because the intervention for CAH had to start so early in embryonic life if it was going to work as planned, it had to begin before doctors could know if the woman was even carrying the kind of embryo they were worried about: a female with CAH. Remember that, on average, only one out of four offspring of at-risk couples will turn out to have CAH, and of those the issue of ambiguous genitalia affects only the females. Bottom line: Seven out of eight of the embryos exposed to dexamethasone in the first trimester could never gain any of the intended benefits from the intervention, because only one of eight "at-risk" pregnancies would produce a female with CAH—but all would bear the unknown risks of early dex exposure.

Actually, when I redid the math after some research, it turned out to be more like nine out of ten fetuses exposed to the risks with no chance of benefit—because not all girls with CAH develop ambiguous genitalia, and because even the proponents of dex estimated that it achieved its goal only about 80 percent of the time. Yet 100 percent of these couples' developing offspring would be blasted, starting in the embryonic stage, in hopes of "saving" 10 percent of them from developing to be sexually atypical. The fetuses identified in utero as females with CAH potentially faced even higher adverse-event risk than the rest, because for them, the dexamethasone dosing continued all the way through the pregnancy, whereas for the rest, the intervention ended after a few weeks, once the obstetrician determined that the fetus being carried was most likely not a female with CAH.

Given all this, even some fairly old-guard doctors were speaking out against this whole intervention, essentially saying "this is just plain nuts." One senior pediatric endocrinologist, Walter Miller of the University of California's medical school in San Francisco, had been warning about the dangers for years. Miller had finally declared in 2008 in *Endocrine News* that "the accumulating evidence that prenatal exposure to dexamethasone has mild but deleterious effects on the developing brain of the 7 of 8 fetuses treated needlessly should clearly indicate that prenatal treatment of CAH is fraught with ethical (and possibly legal) problems. It is this author's opinion that this experimental treatment is not warranted and should not be pursued, even in prospective clinical trials."

But the intervention wasn't even being pursued in prospective clinical trials, except in Sweden, where in 1999 researchers intentionally halted access to it through any other route because of their findings that children had been harmed. As far as I could ascertain, in the United States there had been no meaningful long-term oversight from medical-school institutional review boards (IRBs, or medical ethics committees), so there were no meaningful protections in place to assure that these pregnant women were fully informed before taking on these significant risks. Without IRB protocols in place for a prospective trial that would track the children continuously, without independent safety monitoring, there would be no meaningful assurance that adverse outcomes would be detected and reported promptly.

As I soon discovered, the truly amazing thing was that numerous medical societies had long been calling for the highest scientific and ethical precautions when using prenatal dexamethasone for CAH. Even a committee from the top American pediatrics group, the American Academy of Pediatrics, (AAP), had called for the greatest caution in an open (and ugly) debate with Maria New in 2001, the very year she was recommending the treatment at that support group meeting. The AAP representatives admonished New: "The maxim of 'first do no harm' requires a cautious, long-term approach, which is why the Academy Committee unanimously agrees that prenatal glucocorticoid therapy for CAH should be confined to centers doing controlled prospective, long-term studies. The memory of the tragedies associated with prenatal use of DES (diethylstilbestrol) and thalidomide demands no less."

Yet almost a decade after that admonishment, nothing had changed.

RESOLVED TO WORK on this issue, my only question was *how*. This was different—really different from our old intersex fights over genital surgeries. For one thing, I was actually ideologically *allied* with most of the medical establishment types this time around. For another, intentional exposure of human fetuses to a development-altering steroid was something about which all kinds of protections were *already* supposed to be guaranteed. It seemed we just needed to pull some particular lever. But which one?

Before I did anything, based on what I had learned from talking to criticized researchers, I thought I had best first write to Maria New and ask her whether she had been obtaining informed consent of the parents. Maybe I was wrong. Maybe there was an explanation for what I thought I was seeing. When we had met years earlier at a meeting of intersex specialists, I had walked in to find Maria New recommending my first book to her colleagues, even though it contained sharp ethical criticisms of them all. She surprised me then. Maybe she would surprise me now. So on December 8, 2009, I wrote to her: "Dear Maria, I hope this finds you well. I have two questions for you with regard to prenatal dex treatments for CAH: Do you have any forthcoming publications on the long-term outcomes of these treatments? May I see the consent form you're using for this experimental treatment?"

There came no reply. But it was safe to guess that New saw my message; in early January, at a conference of pediatric endocrinologists in Miami, the clinicians who had called me found her even more defensive than usual. In a session on long-term follow-up of prenatal dex for CAH, New raced through the slides, failing to let anyone really see the data. From what those calling to report to me could see, more than half of those exposed appeared to be completely unaccounted for, and the scientific quality of what data New did purport to have was very poor. One of my colleagues present at the meeting, pediatric geneticist Eric Vilain of UCLA, had boldly pressed New during the Q&A: What did the consent for this look like? What was she telling the mothers? She responded that his question was out of line. She wasn't going to answer him. Others told me that her response to Eric had caused the room to go still. Then she'd gone on to assure the audience that she wasn't accidentally making the prenatally exposed boys gay. Perhaps she said this because she knew that in the Swedish cohort at least one boy exposed in utero had been born with genitals that suggested incomplete masculinization, a sign that perhaps the boys were also becoming collateral damage in the sex-normalization quest.

I couldn't let this continue without trying to help bring it to light. After pulling all the published information I could find and looking for evidence of proper ethics and scientific oversight, what I was seeing just seemed to confirm our worst fears. Besides promoting the intervention via the support group for CAH and her own foundation Web site, when writing about CAH for various textbooks, Dr. New had made a point of plugging prenatal dex to other doctors, writing as if it simply was the standard of care among clinicians in regular practice. As a result, all over the country, obstetricians and genetic counselors were using prenatal dex believing it to be safe and effective.

Still unsure exactly what to do, I talked to Aron's boss, Marsha Rappley, who just happened to be recently retired from chairing an FDA panel on pediatric therapeutics. Marsha seemed skeptical that people would really be giving a fetus-altering drug outside of prospective long-term studies, but she helpfully suggested that, in addition to writing the specific medical schools involved, we might also formally register our concerns with the FDA Office of Pediatric Therapeutics and the federal Office for Human Research Protections (OHRP). *This* was the big idea I needed—to call specifically on the federal agencies charged with protecting people in just these kinds of situations. The same day I talked to Marsha, I also purchased the domain name FetalDex.org as a way to provide information for everyone, and I let the editors at the Hastings Center, the largest independent bioethics center, know I was planning to write a blog post for them about prenatal dex. The article would be used to gather up names of people in bioethics and allied fields willing to sign letters of concern to the FDA and the OHRP.

Still feeling more than a bit queasy, I decided I'd also better write another e-mail message to Maria New, letting her know I was about to start criticizing her publicly and giving her another chance to show me evidence that she was protecting the rights of the families. The subject line I used this time was "IRB oversight," to indicate clearly that I was worried about violations of regulations governing research on humans:

> I am about to write and publish an article critical of your use of dexamethasone in the maternal-fetal period without IRB oversight. You have a major grant to study the long-term effects of this experimental treatment, which clearly indicates you understand that there are risks to the children (not to mention the mothers) subject to what you understand to be an experimental treatment, yet apparently you do not have IRB approval for the experiment that is the treatment itself. In the event you have IRB approval for the experimental, off-label use of dexamethasone in pregnant women "at risk" of carrying a child with 21-Hydroxylase Deficiency [CAH], please let me know that within 48 hours. In the event I am wrong, I will correct my work. Thanks.

Two days later, I got a response, not from Dr. New but from Jeffrey Silverstein, the director of the Program for the Protection of Human Subjects at Mount Sinai, i.e., the person who would be in charge of overseeing New's human-subjects research.

Silverstein wrote to say, "While your email was unusual, I am responding in the spirit of academic collegiality. In the time frame given, I am only able to confirm that Dr. New has an IRB-approved protocol for the study of 'Rare Diseases of Steroid Metabolism,' that includes treatment with dexamethasone in pregnant women with 21-Hydroxylase Deficiency."

I knew from searching ClinicalTrials.gov (the government Web site showing who has what federal clinical research funding) and Dr. New's papers what that probably really meant: Mount Sinai had in all likelihood given permission for the low-risk retrospective follow-up studies on prenatal dex for prevention of sex atypicality, studies conducted years *after* the pregnancies, but not for the high-risk prenatal drug administration itself. If so, there were no ethical protections in place at the important moment—when the woman was offered the drug.

Nevertheless, I asked Dr. Silverstein to clarify: "I know Dr. New has IRB approval for the . . . ongoing study of children who were subjected to dexamethasone in the womb. I am asking whether she has IRB approval for the treatment of pregnant women themselves with dexamethasone, a drug use understood to be experimental and off-label."

He didn't answer.

CHAPTER 8

DOCTOR, MY EYES

IN A STRICT LEGAL SENSE, American medical researchers are not obligated to follow the principles laid out in the Nuremberg Code, the code of ethics drafted in 1947 in response to Nazi medical research atrocities. Nor are they legally bound to adhere to the Declaration of Helsinki, the expanded medical research ethics code adopted by the World Medical Association in 1964. But American medical ethics regulations do require adherence to the same basic principles advanced in these foundational codes.

What was going on with prenatal dexamethasone for CAH seemed to violate a number of those principles, including the first principle of nearly all research ethics codes: Except in extraordinary circumstances, researchers must obtain the voluntary informed consent of a subject to participate in an experiment. There appeared to be a fundamental failure of informed consent in prenatal dex "treatment." New consistently described the intervention to the CAH families as safe *even while* she was telling the NIH she needed grant money "to establish that prenatal treatment with dexamethasone is safe." Did the families New encouraged to use prenatal dex know they were being used as "human subjects of research" for funding purposes? It sure didn't look like it to me.

Medical research ethics codes also called for animal modeling before use of a drug on humans, yet, in spite of the extreme high risks in intentionally changing fetal development, no animal modeling of intersex prevention had occurred. The codes also required that researchers do the best science possible; if researchers are going to put people in harm's way as experimental subjects, ethics require that they design the study to produce the best data ascertainable. Unfortunately, the way Dr. New was "studying" this intervention—primarily using retrospective phone surveys of only a portion of the population exposed—was completely inadequate to ascertaining the risks and harms involved. No placebo-controlled trials, no blinded studies, no prospective studies, and inadequate retrospective studies—the science here was so bad that it itself could be viewed as unethical.

What was going on was so out of line with accepted standards that many of the physicians and ethicists I called for advice on what to do simply could not believe what I was describing. It seemed especially shocking to many that someone of Dr. New's stature would engage in such reckless behavior. Historically, though, it's researchers of the highest stature who get away with this sort of behavior the longest. They have the reputation and the resources to weather criticism. And New had reputation and resources. At Cornell, New had been chair of Pediatrics and the head of the children's clinical research unit at Cornell; as such, she had been one of the most powerful pediatric researchers in the country. Even after she left Cornell, because the fraud case related to New's grant had been settled with no admission of wrongdoing and because by then New had found herself a new job at Mount Sinai School of Medicine, to the outsider there appeared to be no blemish on her record. She boasted membership in the National Academy of Sciences, a history of leadership in her field, the adoration of thousands of CAH-affected families, and a remarkable three decades of continuous NIH funding.

One might wonder how a researcher of Maria New's stature could bring herself to violate the most basic principles of medical research, but if history is any guide, like most American researchers who engage in unethical behavior, Dr. New probably had the attitude shared by many American medical researchers: that the Nuremberg Code was not written for her because it was a code meant for Nazis, not for good people. In public Maria New positioned herself as a brave advocate for CAH-affected families, fighting back against the seemingly unreasonable—even uncaring?—calls for extreme caution from groups like the American Academy of Pediatrics, the Pediatric Endocrine Society, and the European Society for Pædiatric Endocrinology.

Why didn't those groups do more than issue cautions? Medical societies like these cannot discipline individual researchers. The only institutions that can are the researchers' own medical schools and certain official authorities, including the Office for Human Research Protections (OHRP) and the FDA. For them to act, typically someone would have to make a complaint. No clinician had done so, presumably because they all knew the kinds of backlash they'd get for breaking ranks.

In late January 2010, I started using various networks to collect names of academics in bioethics and allied fields who would be willing to sign a formal "letter of concern" about the way prenatal dexamethasone for CAH was being deployed under Maria New's leadership. Early in the process, Ellen Feder, a tenured professor of philosophy at American University, whom I knew from academic intersex studies, offered to sign, and I asked her if she'd do more—draft the letter and become the corresponding author. She did so, creating a letter that would be as much about alerting Cornell, Mount Sinai, the Office for Human Research Protections, and the FDA as alerting obstetricians, genetic counselors, CAH-affected families, and health reporters.

In our formal letter, we questioned using such a risky intervention for a matter of genital appearance. We indicated serious concern that there had been inadequate institutional review board (ethics committee) oversight for the pregnant women, and that there had been inadequate scientific rigor as well. We pointed to Dr. New's misleading promotions of the intervention as safe. We referred to studies that suggested that the use of glucocorticoids in pregnancy could be seriously harmful. The letter ended:

> We call for rigorous investigation into possible regulatory violations in this matter. We also believe that women who have been treated without the protection of IRBs should now be advised of the information that may not have been made available to them at the time of treatment, and that they should be given the most recent information from studies indicating long-term risks to women and children. Finally, we agree with Dr. Miller, Distinguished Professor of Pediatrics and Chief of Endocrinology at the University of California San Francisco, who has written that "this experimental treatment is not warranted and should not be pursued even in prospective clinical trials."

I felt certain that the federal investigation we were calling for—an investigation that would be independent of politics, transparent, evidence-based, and rule bound—would finally produce a statement that everyone, including Maria New, would have to heed. The feds would have to recognize that what had happened *was* another DES—and a deeply troubling case of doublespeak by its main promoter-researcher, too.

By early February, Ellen and I had signatures on the letter from thirty-two academics in bioethics and affiliated fields from a total of twenty-seven institutions, enough for us to feel we should go ahead and submit. The sooner the letters reached the people with real power, the sooner the CAH-affected families might have their rights protected. I also calculated that, with thirty-two signatories, we had a big enough group for reporters to notice. Garnering press attention had to constitute a major strategic aim. We needed external eyes and ears if we were going to make it impossible for Maria New to work against all the doctors calling for ethics and science in this case. Besides, government paper pushers were more likely to act if journalists came asking what they were doing about a problem involving pregnant women and experiments on fetuses. And it was just that hook—experimentation on pregnant women and fetuses—that made me sure it was going to be easy to get reporters interested in this story.

But once I tried calling reporters about dex, I began to understand just how much the tradition of investigative journalism had been weakened in the decade since the early intersex rights movement by the economic struggles of magazines and newspapers in the wake of the Internet's rise. Previously, knowledgeable and engaged investigative health reporters had been plentiful and critically important to the success of our work. Now, a number of reporters I tried to contact from my address book seemed to have left the business. A few of the national health reporters I did reach told me the story was too complex to cover, given the short time frames their editors could allow them nowadays to work up a single story. I also got the sense that some reporters didn't like the idea of having one particular researcher named—and for that hesitation, knowing what had happened to various researchers vilified for ethical violations and/or bad science when the press rushed to repeat accusations, I would not have been so troubled if I hadn't had the sense that this hesitation had less to do with fear of smearing someone than a newfound editorial fear of costly fact-checkers and lawyers. The American press, once a large and fearsome institution, now seemed emaciated and timid, jumpy and distracted.

All we could do was push on. Not long after we submitted the letters, staff members from the OHRP and the FDA had let us know they would be looking into the matter—a development that made me both hopeful and genuinely sick to my stomach. Strange as this may sound, until I read these replies, I hadn't really specifically envisioned what my actions might do to Maria New as a person. I certainly didn't want to ruin her or her reputation; I just wanted the CAH-affected families' rights protected, and her and all the clinicians involved to fix all the ethical and scientific problems in this situation. Now I found myself imagining the octogenarian New going through the whole descent into blackness that I had charted in so many researchers accused of being unethical. At least this left me feeling less disappointed in the lack of press interest; perhaps it was best that this stay out of the press until an independent group of government investigators confirmed what we and her other critics were seeing. Not that there was any question in my mind that the government would confirm that what New had done was deeply unethical. But there was also no question in my mind that Dr. New had bounded ahead thinking she was doing what was best for these families. That's exactly why she had been so persuasive: She had all the charisma of the true believer.

As the weeks went by, I found myself having to go for more midday runs than I usually do in the winter. In spite of finding more and more evidence that what had happened had been both unscientific and unethical, I wasn't happy to be caught up in yet another tangle. If it hadn't been for my friends calling me in to help with a rights movement that had so long been dear to me, I would have ignored this as I had so many recent invitations to investigate other controversies. To make matters still more uncomfortable, in contrast to our early intersex rights work—when no surgeon would go on record to defend cosmetic genital surgeries on intersex children—the work on dex had quickly started to generate pushback. Answering us at the Hastings Center's blog site, one group of Boston clinicians, specialists in intersex treatment, claimed that we were misrepresenting the problem as a "cosmetic" treatment. They insisted that dex was critically necessary to prevent the urogenital sinus, the condition in which, in some CAH females, the urethra and the vagina join, increasing risk of infections and problems with vaginal intercourse. But as we said in the response I rapidly drafted, in the literature on dex, New (like other researchers) didn't talk about it as primarily being a preventative for urogenital sinus. In fact, when writing about dex, physician-researchers barely talked about the urogenital sinus *at all*. The specialist medical literature made clear that the primary goal of dex was prevention of atypical genitals in genetic females. The goal was to give the parents hope of a "normal" girl. New's major critic on dex, the pediatric endocrinologist Walter Miller, acknowledged this truth dryly: "It seems to me that the main point of prenatal therapy [for CAH] is to allay parental anxiety. In that construct, one must question the ethics of using the fetus as a reagent to treat the parent, especially when the risks are non-trivial."

Meanwhile, on a major bioethics e-mail discussion list, we were being sharply criticized by a bioethicist out of Texas's Baylor College of Medicine, a fellow named Larry McCullough. According to McCullough, we had utterly misrepresented the question of IRB oversight; Dr. New, he claimed, had had proper oversight. Our letters of concern were full of gas, declared McCullough, a man who had mastered the authoritative medical-expert tone. This had several of my collaborators sweating bullets until I rechecked the papers to which McCullough was referring. Doing so confirmed what I'd seen before: In the 2001 "extensive personal experience" report on 532 pregnancies subject to this intervention, Dr. New appeared to have IRB approval only for subsidiary studies, like whether a particular genetic screening technology picked up CAH, or for retrospective phone surveys, *not* approval for the prenatal drug intervention itself. And it was the drug intervention that was the real issue. I wrote an evidence-based response to McCullough, and sent it to the list.

All this angry pushback had me feeling like I was chewing tinfoil for gum. Had I done this wrong? I hadn't taken the time to do anything like a full-blown historical investigation on this—nothing like the year I had spent on Bailey's controversy or the year on the Tierney fiasco. Of course, that was in large part because it all seemed pretty clear and very urgent. Moreover, much of what we really needed to know to judge the scene—i.e., what had really happened at Cornell and Mount Sinai in Dr. New's clinics—could in all likelihood only ever be found out via a formal governmental inquiry.

But maybe I should have done even more research before setting out as I had. I had dragged in so many colleagues who had basically signed on my word. Or more specifically, on my reputation for being meticulous. . . . I kept doing this self-check, to see if the problem was that I had done this because I had wanted to prove something—something that might blind me to the facts. Had I become self-deluded? Even self-righteous? But I was pretty sure self-righteous people didn't feel this anxious, this unsure of their own actions. Still, at my request, Aron and Ellen kept going over the evidence with me. And I kept reminding myself: The reason those good clinicians had called me was because all the conventional routes had been tried and had failed.

It helped that just a few weeks after our letters went to the feds, a clinical insider leaked me a draft of a forthcoming new medical consensus on prenatal dexamethasone for CAH. Besides acknowledging that "the mechanism of dexamethasone's action in the fetus is incompletely understood" and confirming that high-quality efficacy and long-term safety data were virtually nonexistent twenty-five years into the practice, the authors admitted that "the condition being treated, while fraught with emotional complexities, is *directed toward a cosmetic outcome* rather than aiming to preserve life or intellectual capacity." (Prevention of the urogenital sinus as the true goal, my ass.) Given all this, the authors concluded, "We recommend that prenatal therapy continue to be regarded as experimental, and be pursued only through research protocols approved by the local Institutional Review Boards at centers capable of enrolling a sufficiently large number of participants in such protocols—alone or in collaboration with other centers—to yield precise findings."

Here was yet another major medical attempt to stop promotion of this high-risk game as standard care. And this was no minority report; the document had been put together (and put together *before* we ever started making noise) by the Endocrine Society, a group Dr. New had for decades helped lead. The cosponsors included the American Academy of Pediatrics, three major pediatric specialist groups, and even the support group that had long been New's supporter and promotion venue.

Of course, it being a leak of a draft consensus, I didn't dare share this beyond the closest insiders. I worried that our activism might paradoxically cause the consensus group to pull back in the published version, wanting to distance themselves from the people who had sicced the Feds on a fellow physician. If there's one intervention almost all specialists recommend, it's closing ranks against activists.

All I could do was wait and hope the government workers on whose plate this had landed would be moved to action. I kept picturing them as accountants on white horses.

IN MAY, on one of the first warm days of spring, I checked my e-mail just after a lazy outdoor lunch with a friend and found that my in-box had exploded with the same message forwarded to me by colleagues all over the country. I opened and read one. My heart starting pounding like timpani in my left ear. The *American Journal of Bioethics* had just sent out a formal call for responses to a new "target article" slamming our call for a federal investigation on prenatal dex. The article was called "A Case Study in Unethical Transgressive Bioethics: FetalDex.org's Letter of Concern about the Prenatal Administration of Dexamethasone." (FetalDex.org was the public Web domain I'd set up to be transparent about our efforts and to educate families, reporters, and doctors about what was known and unknown about prenatal dex.) Reading the abstract made it feel like a fast swarm of bees was entering my head through my eye sockets:

> On February 3, 2010, a "Letter of Concern from Bioethicists" was sent by fetaldex.org to report suspected violations of the ethics of human subjects research in the off-label use of dexamethasone during pregnancy by Dr. Maria New. . . . We provide a critical appraisal of the Letter of Concern and show that it makes false claims, misrepresents scientific publications and websites, fails to meet standards of evidence-based reasoning, makes undocumented claims, treats as settled matters what are, instead, ongoing controversies, offers "mere

> opinion" as a substitute for argument, and makes contradictory claims. The Letter of Concern is a case study in unethical transgressive bioethics. We call on fetaldex.org to withdraw the Letter and co-signatories to withdraw their approval of it.

I quickly downloaded the full manuscript, and could easily see that these charges made against our group were absurd. Chief among the accusations was the contention that we'd made our complaint to the Feds without performing a systematic review of the literature—the type of major scientific review that is rightfully conducted by epidemiologists, not ethicists. The truth was that, even without such formalized epidemiological study, we and the critics before us had *plenty* of evidence-based reasons to raise concerns about what Maria New had been doing with these families, much of it from Dr. New's own pen. The target article's authors also objected that we cited only the array of research that suggested potential harm from prenatal dex, and not New's group's more optimistic literature, as if the presence of a lively debate between New and her large number of physician-critics negated all the ethical and scientific problems. (Translation: "If something is controversial, then there is no truth.")

In fact, the target article's authors used a 2002 medical society position paper on the intervention to show "that there is substantial difference of opinion concerning whether prenatal treatment of CAH is a research endeavor." This was supposed to prove that doctors varied widely in their opinions about prenatal dex—that some might agree with New, and some might disagree. What the target article's authors didn't tell readers is that that 2002 position paper—a joint statement from the European and American pediatric endocrinology societies—actually concluded this:

> We believe that this specialized and demanding therapy should be undertaken by designated teams using nationally or multinationally approved protocols, subject to institutional

> review boards or ethics committees in recognized centers. Written informed consent must be obtained after the balanced review of the risks and benefits of treatment. Families and clinicians should be obligated to undertake prospective follow-up of prenatally treated children whether they have CAH or not.

In other words, the pediatricians signing on to this position statement in 2002 weren't, as the target article authors implied, saying, "This is controversial"; they were essentially saying, "Use prenatal dexamethasone for CAH only as the Swedish group has, not as Maria New has."

The target article authors went on to insist that because New told them that she personally had written only one prescription for prenatal dexamethasone for a pregnant woman—because allegedly "the referring obstetrician obtains consent and writes the prescription"—we were wrong to go after her. But as everyone involved knew, Maria New openly and personally took credit in her online clinic advertisements for having "treated" over six hundred women with prenatal dex. (In the 2001 presentation of which I had a recording, she had told the CAH-affected families that her Cornell clinic had "conducted [more] prenatal diagnosis and treatment . . . than everybody else in the world put together.") In another instance of diversion from the real issues, the authors assured readers that New had IRB approval for the follow-up studies that she was doing years after prenatal exposure. To reiterate, those retrospective follow-up studies were never our concern. Our concern had always been that the women didn't benefit from IRB oversight *when they were pregnant*, when they would have had a chance to give or refuse informed consent to the drug experiment. Most bizarrely, the term "transgressive bioethics" was being used to signify that we had wandered out of bioethical theory and actually done something to defend patients' rights. How depressed should we be, I asked Ellen, when standing up for patients' rights makes us "transgressive" as ethicists?

The lead author of this target article was none other than Larry

McCullough, the Texan bioethicist who had been fiercely criticizing us on the bioethics e-mail list. Why was this guy so doggedly coming at us? The name of the article's second author, Frank Chervenak, suggested an answer. Chervenak, as I soon learned, was McCullough's lifetime collaborator; they had by that point published over 150 ethics articles together. What was Chervenak's professional position? Chair of Obstetrics/Gynecology and director of Maternal-Fetal Medicine at . . . wait for it . . . Weill Cornell Medical College, where over six hundred dex interventions had apparently occurred when New worked there—almost certainly, as New claimed, far more than anywhere else in the world. Without ever saying so explicitly in the paper, Chervenak was obviously defending his own institution against our calls that Cornell be investigated, and presumably defending the department *he was in charge of.* But the average reader of the paper simply wouldn't pick up on this without an explicit disclosure, and the paper included no such thing. McCullough, Chervenak, and two colleagues had produced a paper that looked like an objective accounting of how incredibly wrong we were. It was, in fact, crafted to write us off as an activist group and to show that activists are unscientific and uninterested in facts.

My head was killing me. It wasn't that I couldn't take a debate. What upset me was the fact that this article would carry the legitimacy of real scholarship and it would be indexed in the medical literature—meaning that all the people who had trusted my word when they signed the letters would be labeled unethical and intellectually incompetent in a "peer-reviewed" article. The lay reader would have *no idea* about the reality behind the *American Journal of Bioethics* (*AJOB*), the journal publishing this. I knew the truth about *AJOB* and how it operated. It was founded and still run with a nearly complete lack of transparency by Glenn McGee, a man whose behavior was so questionable he had become the subject of a six-page spread in *Scientific American* entitled "An Unethical Ethicist?."

But who would ever suspect that the *American Journal of Bioethics* might be fundamentally unscholarly—even unethical—in its publishing practices? What were the odds the government investigators would see *AJOB* and this article for what they were? More likely they could just decide that our "peers" had judged our actions "unethical" and "transgressive." No doubt that was the goal of this article—to undermine the investigation, to conclude this was a fight between some feminist versus establishment bioethicists.

Sure that I'd lose, I did try calling foul. I wrote to McCullough to tell him to correct the manuscript so that it did not confuse the Web site FetalDex.org with the group of academics who signed the letters of concern. I asked McGee to move to correct the paper's inaccuracies and also asked him to share with me whatever conflicts of interest McCullough and Chervenak had disclosed. In response, McCullough claimed to me that his group had no meaningful conflicts of interest to disclose. Right.

I tried to keep my wits, but here was the thing ringing around in my head: I didn't personally know most of the people who signed the letter. With McCullough et al. now calling them unethical and transgressive—would some disavow our letters? Say I'd misled them? Were these guys going to not only undo the federal investigation, but ruin my reputation in the process?

I hadn't come here to defend myself. I mean, at the superficial level, it seemed reasonable for them to come after me; I myself had said in my work on controversies that one should always examine the accuser as much as the accused. But with the way McCullough and company were framing the discussion, I was starting to fear I had achieved exactly the *opposite* of what I'd hoped when I'd dreamed of using this case to inspire others to do this kind of evidence-focused work to stop truly unethical research practices on truly vulnerable people. Instead of showing a way to behave like a scholar who also cares about social justice, I had attracted criticism so venomous, it might well scare other academics completely off the path of advocacy. It might simply encourage people to believe there *is* a hopeless division between those who truly care about science and those who claim to care about social justice.

As IF THINGS were not complicated enough, another problem suddenly appeared on my radar. Because of the dex work, I had been back in close contact with an old intersex-activist colleague, Janet Green, a middle-aged patient advocate born with CAH. In that late spring of 2010, Janet said to me, "Listen, I'm thrilled you're back and working on dex, but I hope when you're done with that, you'll turn your attention to the vibrators." I joked that I was always happy to turn my attention to a vibrator, but then asked what she was talking about. She explained that she was referring to work being done by Dr. Dix Poppas, the pediatric urologist at Weill-Cornell who happened to be Maria New's favorite surgeon for clitoroplasties, including clitoral-reduction surgeries. Basically Poppas was taking girls, aged six and up, whose clitorises he'd earlier surgically shortened for psychosocial reasons, and having them come in for a follow-up exam involving what he called "clitoral sensory testing" and "vibratory sensory testing." He'd make them lie back on an exam table, fully conscious, while he brushed these little girls' inner thighs, labia majora, labia minora, vaginal openings, and clitorises with a cotton-tip swab. While he did this to them, he'd ask them to report on a scale of 0 to 5 how well they could feel his stroking. Then he'd do the same with a medical vibratory device: vibrate the little girl's inner thigh, vibrate her labia, vibrate her clit. Janet pointed me to the paper where all this was published, in the *Journal of Urology*, in 2007.

As I read the paper, astonished, I kept thinking about one thing: the place where Poppas performed these surgeries and his follow-up stimulations was the New York–Presbyterian Hospital—the very hospital where Poppas's predecessors had amputated Bo's clitoris. Not sure what to do, I called a phone meeting of Janet, Ellen, and Anne Tamar-Mattis, JD, a California-based legal advocate for intersex rights and for several months now a key colleague in the prenatal dex work. (Anne had sent her own legal analyses to the Feds to back up our letters and had also arranged a group letter from intersex adults.) As we talked, we had no doubt we would have plenty of company in feeling outraged. Like New, Poppas also seemed to be violating key research regulations; he had no advance IRB approval to be doing these examinations, even though they seemed quite clearly designed to collect data—data he used to respond to critics like us who said that his clitoroplasties were of high risk to a girl's sexual sensation.

After some discussion, I wrote with Ellen another piece for the Hastings Center's online *Bioethics Forum*, this one called "Bad Vibrations." Anne prepared legal letters to the various individuals charged with protecting subjects at Cornell and at the federal level. And I called Dan Savage, the sex-advice writer and LGBT advocate, and asked him to help us out. Although I now understood that investigative health journalism had, thanks to the Internet, shrunk like an ice cube in the sun, I knew a prominent pundit like Dan could position this story to really matter nationally. In our phone call, Dan made clear he was fully committed to doing so. I explained to him that we needed this one to keep leveraging the dex story as well, to keep it from getting buried as an internal controversy and put away, now that McCullough et al. had marked us as not worth believing. Various journalists had started to look into dex, but none of them had published on it. Maybe with this, with a *pattern* of abuse of this population at Cornell, we might get some journalistic light shone.

On June 16, 2010—right about the middle of the calendar year that I still was thinking couldn't get much worse—*Bioethics Forum* published "Bad Vibrations," and Dan Savage pushed it hard for us. None of us expected what followed: The blogosphere exploded with the news that culturally motivated female genital cutting wasn't just something that happened in Africa, it was going on in the United States—this in spite of the fact that these surgeries were something I and many others had been talking about for the last fifteen years!

But as I had suspected might happen in the great Rube Goldberg machine of the American media, the noise about Dix Poppas's scalpels and vibrators led *Time* magazine's editors to finally publish an "unrelated" article they'd been sitting on, a terrific piece of reporting on prenatal dex for CAH by a reporter named Catherine Elton. (OK, I'm sure it didn't hurt that I contacted Elton to tell her about the Poppas kerfuffle, hinting she might want to tell her editors to piggyback on it.) The *Time* article came out only two days after "Bad Vibrations," and it began:

> When Marisa Langford found out she was pregnant again, she called Dr. Maria New, a total stranger, before calling her own mother. New, a prominent pediatric endocrinologist and researcher at Mount Sinai Medical Center in New York City, is one of the world's foremost experts in congenital adrenal hyperplasia. . . . "Dr. New told me I had to start taking dexamethasone immediately," says Langford, 30, who lives in Tampa. "We felt very confident in someone of her stature and that what was she was telling us was the right thing to do." . . . Langford says also that neither New nor her prescribing physician mentioned that prenatal dexamethasone treatment is an off-label use of the drug (an application for which it was not specifically approved by the government) or that the medical community is sharply divided over whether dexamethasone should be used during pregnancy at all.

Elton's *Time* report also relayed the story of a twenty-four-year-old mother named Jenny Westphal "who took dexamethasone throughout her pregnancy at the recommendation of another doctor" and who "feels misled. Like Langford, she was not asked to give informed consent. Unlike Langford, however, her daughter, now 3, who has CAH, has also had serious and mysterious health problems since birth, including feeding disorders, that are not commonly associated with her adrenal-gland disorder." Westphal had been looking online, and had apparently heard our alarms, just as I had hoped the exposed mothers would. (The reporter Elton had found these mothers on Listservs where they were talking about our call for a federal investigation.) "I was outraged, frustrated, and confused," Westphal told *Time*. "Confused, because no one had ever warned me about this. I wasn't given the chance to decide for myself, based on the risks and benefits, if I wanted the treatment or not."

Then the very next day came the news, in the specialty paper *Endo Daily*, that the big medical consensus group had finally announced its stance on prenatal dexamethasone for CAH: "Prenatal treatment of CAH [should] continue to be regarded as an experimental procedure. Prenatal treatment should not be conducted on a routine basis but should be confined to formal clinical study situations under institutional review board protocol." The article went on to say that "the task force was hampered by the lack of high-quality data. Of 1,083 'candidate studies' originally identified, only four met the quality criteria [for scientific reliability] agreed upon by the sponsoring groups" for further analysis. (Two were from the Swedish group, two from New's group.) The sparse, scientifically weak data that did exist suggested "modest improvement" in genital outcomes, and "side effects" that included stillbirth and malformation. The task force chair, a physician who had once worked with New at Cornell, told *Endo Daily* bluntly, "This is not standard of care." I alerted the signatories to our letter, to try to bolster their confidence that we had done the right thing.

FOR SOME TIME, Anne, Janet, and Ellen had been noting to me the extent to which for Maria New, the goal of medical intervention in these cases wasn't just a girl who looked good, but one who behaved according to New's notion of femininity. New was joined by her close collaborator, the Columbia University psychologist Heino Meyer-Bahlburg, in her focus on gendered behaviors in females with CAH. In a 1999 paper entitled "What Causes Low Rates of Child-Bearing in Congenital Adrenal Hyperplasia?" Meyer-Bahlburg had put it this way:

> CAH women as a group have a lower interest than [non-CAH-affected] controls in getting married and performing the traditional child-care/housewife role. As children, they show an unusually low interest in engaging in maternal play with baby dolls, and their interest in caring for infants, the frequency of daydreams or fantasies of pregnancy and motherhood, or the expressed wish of experiencing pregnancy and having children of their own appear to be relatively low in all age groups.

WHAT COULD BE DONE about CAH-affected women's "unusually low interest" in being mommies? The end of Meyer-Bahlburg's article gave a clue: "Long term follow-up studies of the behavioral outcome will show whether [prenatal] dexamethasone treatment also prevents the effects of prenatal androgens on brain and behavior."

In a similar vein, in 2010, Maria New coauthored an article in the *Annals of the New York Academy of Sciences* that construed a lack of interest in babies, in "women's" activities, and in having sex with men as being "abnormal" for women—abnormal enough to justify attempts at prevention with prenatal dexamethasone: "Gender-related behaviors, namely childhood play, peer association, career and leisure time preferences in adolescence and adulthood, maternalism [interest in becoming a mother], aggression, and sexual orientation become masculinized in 46,XX girls and women with 21OHD deficiency [CAH]. These abnormalities have been attributed to the effects of excessive prenatal androgen levels on the sexual differentiation of the brain and later on behavior." New and her coauthor concluded, "We anticipate that prenatal dexamethasone therapy will reduce the well-documented behavioral masculinization."

As I was slow to figure out, the interest by New and her collaborators in whether one could prevent tomboyism and lesbianism with prenatal dex actually had been an open secret within the specialist medical scene. My friend the Michigan pediatric psychologist David Sandberg had spoken to this in the *Time* magazine article when he said that New's presentations could give "clinicians the idea that the treatment goal is normalizing behavior." He had added, "To say you want a girl to be less masculine is not a reasonable goal of clinical care."

Just after the *Time* article on dex emerged, figuring we had gotten people to start noticing the apparent research abuses and that it was time to go public about this problematic political facet of prenatal dex, Ellen, Anne, and I put together another essay for *Bioethics Forum*, showing that New and her collaborators had been seeking to see if prenatal sex could lower rates of tomboyism and lesbianism. At my request, Dan Savage again helped us out. Now the media beyond *Time* finally went wild on prenatal dex. But what did the story become? The story of "the anti-lesbian drug." Anybody could write passionate commentary on that. Debating the prevention of homosexuality didn't take any actual work to understand or explain things like FDA protections and institutional review boards—the stuff we most cared about in terms of the CAH-affected families' rights. I went from chewing tinfoil to chewing tacks.

On and on this went, and I became ever more selfishly regretful at having pulled the lid off this can of worms. My life seemed stalled in this uncomfortable mess, and there seemed to be no resolution in sight. Then all of a sudden, seven months after we had sent off our letters of concern, resolution of a sort *did* arrive. In early September 2010, in a pair of short memos, the OHRP and the FDA indicated they had done some poking around.

And they had found nothing worth pursuing.

OHRP found that Dr. New had enrolled some pregnant women in IRB-approved trials years back at Cornell (they didn't say how many of the hundreds she was claiming to have "treated" through her clinic), and she had IRB approval now for the retrospective studies at Mount Sinai. The OHRP reviewers also accepted Dr. New's defense that at Mount Sinai she wasn't the one actually writing the prescriptions. Implication: How could she be held responsible?

For his part, the FDA investigator revealed that in 1996 the FDA had given Dr. New an "investigative new drug" exemption, which he claimed had been "for the administration of dexamethasone during pregnancy for the purpose of preventing virilization in females with congenital adrenal hyperplasia." If this was true, it meant that the FDA had basically decided in 1996 that the fetal experiment was not special enough to warrant extensive FDA review. This seemed astonishing—the FDA, where Francis Kelsey had stopped thalidomide, thought this intentional fetal experiment hadn't warranted careful FDA review? So the investigator claimed. Beyond this, he simply noted the debate over prenatal dex in the literature, said more studies would be helpful, and reiterated that doctors can use drugs off-label if they believe the benefits justify the use.

Maria New wasn't doing anything actively prohibited by regulation, in the estimation of the federal investigators. The OHRP and FDA concluded that we had no case.

Within about twenty-four hours of the feds' announcement, *AJOB* officially issued the published version of the McCullough and Chervenak paper. Of course, the "coincidental" timing amplified by several orders of magnitude the claims made in the *AJOB* target article. The fallout was pretty predictable. The Kansas City–based Center for Practical Bioethics, new home of *AJOB*'s editor-in-chief Glenn McGee, hosted a triumphant podcast with Larry McCullough essentially declaring victory over us unethical transgressive types and plugging the latest issue of *AJOB*. For his part, Maria New's chief collaborator in dex, Heino Meyer-Bahlburg, announced the Feds' nonfindings along with the *AJOB* issue on the big sex-research discussion list, the one that happens to be run out of Northwestern by Mike Bailey. My in-box flooded with messages. I read them all as *you were so wrong*.

CHAPTER 9

DOOMED TO REPEAT?

I HAD BEEN EXCITED by the idea of asking the FDA and OHRP to help with prenatal dex not only because I believed those agencies would have the power to do something the medical societies couldn't, but also because I had imagined impartial and well-informed investigators—those accountants on white horses. Hard as I had tried to look for evidence that our read of the situation might be wrong, I could never be completely objective. A team of government investigators, on the other hand—they would go in as the ultimate scholars, without allegiances, conflicts of interest, loyalties, agendas, or relationships that might complicate their thought patterns.

Because I had believed this, I had told myself that, whatever the agencies found, I would simply accept it. If they found we were right, well, that would be very satisfying, because then rigorous scientific practices might finally begin around this intervention. Researchers—including Maria New—would know that freedom of inquiry entails a responsibility for just research practices. The CAH-affected families might finally have their rights respected, especially the right not to be subject to experimentation without their consent, and other researchers watching might be more inclined to stick to the rules about informed consent. If the government investigators looked at the facts and somehow found we were wrong, well, accepting that judgment would show that we were putting ourselves second to the truth, the universe was functioning right, and we were behaving ethically.

Nevertheless, I never really believed they *could* find that there was nothing wrong with this picture. How could so many people, so many medical practitioners and researchers, looking at it and being outraged, be completely wrong about what seemed so obviously to be an ethical travesty befalling these families? Even if you didn't care about intersex rights the way I did, you had to see how wrong this was. Yet so the government had ruled: we had no case. For someone like me, who had journeyed a long way to come to the firm belief that following the evidence is the most important ethical principle in democracy, the government's findings meant that not only had I been wrong, but perhaps I had behaved unethically.

Of course, that's what the *AJOB* guys, especially Larry McCullough, were saying everywhere they could. One might think that, having studied by now so many researchers who'd been accused on the national stage of unethical transgressions, I might have had some sense of how to manage my own condition in the days just after the one-two punch of *AJOB* and the feds' nonfinding. Instead I could only watch myself experience the progression of emotions brought about by being labeled an unethical researcher: feeling by turns disgusted with myself and utterly wronged, alternately hiding and lashing out, having a sense of mental house arrest, and general catastrophizing. Mostly I felt as if I had utterly failed the families involved. On that shocking day in early September when the Feds and *AJOB* aligned to write us off, it was as if something cracked in my brain. Over the next few weeks, it felt like the crack just kept getting wider.

Well, when you get what feels like a diagnosis of terminal cancer, it helps to have friends who point out that you're not dead *yet*. That job fell to Aron, Ellen, and Anne. The main thing they did was to point out over and over again, until it clicked, that while the government said they didn't find anything worth pursuing according to *their* investigation and a narrow reading of the regulations, that didn't mean we'd actually been wrong about *the ethics*. It certainly didn't mean we were wrong to do what we had done to try to protect these families' rights and health. Aron asked me this: How often in the history of major medical ethical travesties has the government, when called upon to act, done the right thing the first time around? I couldn't think of a single case. Some people in the FDA had given Frances Kelsey a hard time on thalidomide, and it had taken *thirty years* after the finding that DES causes cancer in prenatally exposed humans for the FDA to withdraw approval of its use in humans—and then only because the drugmaker had finally *asked* the FDA to do so. Maybe government agencies weren't superhuman after all. Maybe I had just been incredibly naive.

Aron also hinted that, when trying to understand where to go from here, instead of looking to stories like Chagnon's and Bailey's—wherein the odds had been long that the truth would ever get sorted out in public—I should look instead to what I had learned in my research from people who had persevered in their work on their own in spite of being accused of being enemies of the people. As a result, although I had become a science historian partly out of my adoration of Stephen Jay Gould, the arch-critic of sociobiology, now I found myself looking to Gould's nemesis, Harvard naturalist Edward O. Wilson, for professional and psychological guidance. The year before the dex disaster, when I was researching Napoleon Chagnon's story, Nap had bugged me to talk to Wilson, both because of how Wilson had helped Nap get through the Tierney tunnel, and because of what Wilson himself had been through for his vigorous defense of sociobiology, a field that had greatly offended many on the left. I still had my notes from a phone interview with Wilson in which he had talked to me about how he'd weathered being maligned and kept his focus on the big picture of his scholarship, scholarship aimed at helping lay people understand themselves.

To be honest, before I talked to Wilson that day for nearly two hours, it had never occurred to me that he could have really suffered.

He seemed to me too famous, too stately, too smart, too polished to have ever had his feathers seriously ruffled by anyone. Yet when I spoke with him, Wilson movingly described what it felt like to be repeatedly, publicly accused by his Harvard colleagues, Richard Lewontin and Gould, of promoting a dangerous right-wing science in his quest to help humans understand their own nature. Lewontin and Gould denounced Wilson's work on sociobiology, for example accusing him in a group open letter to the *New York Review of Books* of "join[ing] the long parade of biological determinists whose work has served to buttress the institutions of their society by exonerating them from responsibility for social problems." This letter made clear what Wilson's kind of biological determinism had led to historically: anti-immigration laws, sterilization laws, and ultimately "the eugenics policies which led to the establishment of gas chambers in Nazi Germany." At Harvard, positioning themselves under the banner of Science for the People, Gould and Lewontin, along with lesser-known scientific critics, had opposed Wilson's alleged science against the people. A group called the Committee Against Racism (CAR) aggressively protested Wilson's "racist lies" on the Harvard campus. In 1977, riffing on the criticisms of Gould and Lewontin, a CAR flyer warned, "Sociobiology, by encouraging biological and genetic explanations for racism, war and genocide, exonerates and protects the groups and individuals who have carried out and benefitted from these monstrous crimes."

Before talking to Wilson, I'd thought it couldn't have been so bad before the Internet. But of course Lewontin and Gould were famous and beloved by the Left. They could rapidly publish almost anything they wanted about Wilson and his science, and with fewer venues for public debate, each of these attacks had a big impact. As I discussed with him how easily an individual's research and public identity could be distorted, Wilson told me he could relate: "Gould and Lewontin could do something like that, change your identity to evil."

I had asked Wilson if there had ever been a group of colleagues who had actively defended him at Harvard. He answered:

> No. Unfortunately, no. In fact, I had several close friends in the Organismic and Evolutionary Biology Department who would sometimes speak to me and just say they were sorry it was happening. One was my close colleague Bert Hölldobler. We were working all the time together in the lab and field. We would spend long periods of time talking about it and trying to figure it out, trying to understand Lewontin. He gave me his unstinting support, and made it clear to Lewontin and others. But no one else said anything or did anything. I think they kind of weren't one hundred percent sure about me. They thought maybe there was something to what Lewontin or Gould were saying. No one [except Hölldobler] ever offered any sympathy or any kind of help. I know at one point Lewontin had unleashed some outrageous statement and [Harvard biologist] Ernst Mayr read it and he spoke to Lewontin and asked, "Why are you doing this? Why are you attacking Ed all the time?" And the response Mayr reported to me was that Lewontin said in effect, well, it's just my nature or personality.

I asked Wilson to share the advice he would give to a younger scholar caught in his position, knowing one's motives and research were being widely distorted in the public sphere. At the time, I had no idea I'd be finding solace in his response only a year later. He answered:

> I think I would tell him or her to ignore it. Pay attention, I mean, and respond if there is some really scurrilous thing being said. But, as much as possible, ignore it, and keep working. And you'll win in the end. I know it isn't easy during fights. I always said to myself, "Don't get into a pissing contest with a skunk." Looking back, if you ask me, what I most resent about all that period, I think the answer, the older I get, the farther behind this gets, as it recedes, I must resent the amount

> of time I wasted. I spent countless hours talking with journalists writing stories about this. They'd come to me and say, "Well, Professor Lewontin just said so-and-so, Professor Gould just said so-and-so." Or, "I've read in the latest thing that they've said this. What do you say to that?" they'd ask me. I couldn't sit by and let them say something that was in fact declaring me a racist and a proto-Nazi. I couldn't say, "No comment." I just wasted enormous amounts of energy and just pure time I could have used for something much more valuable. So my advice would be, this too shall pass. Ignore it as much as you can. Conduct yourself with dignity and with courtesy and let it pass.

Looking back at this interview with Wilson, hoping that I would someday again feel capable of helping people through my work, I drafted marching orders for myself: Stop thinking the press will get it right. Stop thinking you need them to get it right. Keep working for the long term. Disengage from the immediate fight and focus on knowing more, knowing as much as you can.

But in the short term, how to work past my sense of having failed at protecting the rights of families affected by CAH and intersex? For that, I found a possible answer in the survival advice I'd also collected the year before dex from the famous psychology researcher Elizabeth Loftus. Today Loftus is generally recognized as having been right—that human memory is fallible and that you can actually implant false "memories" in a person. However, in the late 1990s, Loftus had found herself in very deep trouble over her work challenging recovery of repressed memories of alleged child sexual abuse. She and a colleague, Melvin Guyer of the University of Michigan, had decided to research the index case of recovered memory. They had ended up gathering evidence that suggested that the poster child for repressed memory of childhood sexual abuse, a woman identified in the 1997 case study as Jane Doe, almost certainly did not experience the childhood sexual abuse at the hands of her mother that had been described in the context of an ugly custody battle. Loftus and Guyer had used public documents and interviews to make their case that the girl's "memories" of abuse likely resulted from suggestions made to Jane Doe when she was a child, not from actual events.

Before Loftus and Guyer could publish, Jane Doe—whose real name is Nicole Taus—made formal complaints, saying her privacy was being violated. As one might expect, people who believed they had recovered memories and psychiatrists engaged in memory recovery angrily rallied against Loftus and Guyer. Eventually cleared by their universities, the two went on to publish their exposé, but Taus proceeded to litigation. Given that Loftus and Guyer had employed conventional investigative journalistic methods and had not even published Taus's real name you would think the case would have been quickly dismissed. But as the claims and counterclaims became ever more complex, the suit survived, making it all the way to the California Supreme Court—and taking over a great chunk of the researchers' lives.

In the end, they prevailed. Because Loftus's story had already been so well documented, what I had wanted to know most from Loftus when we had talked was how she had survived and continued to advocate for those wrongly accused through allegedly recovered memories. Now in my dex mess, I had advice from her to make my own. Some of her strategies seemed obvious: You keep working, you pay good lawyers, and you hold trustworthy friends and colleagues near. But one of her strategies had surprised me: She told me that during the worst years, she took up the habit of watching the Lifetime television network. She explained that the same basic story in different guises runs over and over again on Lifetime, the story of a woman facing tremendous adversity who somehow sticks it out and survives. Loftus's approach made her philosophy clear: If you think you're working for the greater good, you take the knocks and keep working, doing good research to figure out reality. You stop worrying about yourself. And so—staying firmly focused on the work that matters—you survive.

Knowing there probably wasn't going to be a good historian who would do the work to tell me what actually happened with the whole prenatal dex scene, I realized that what Aron was telling me was right: I had to stop acting like a beaten dog and act like a professional historian. I had to treat this like a new major historical project, and I had to have the patience to work on it as long as it took to figure it out. As I made findings, I would have to present and publish them in scholarly arenas and in the mainstream press, so that people could see the evidence I'd found and help bring justice to bear. It might take several years, but this project was important enough to give years, because this really did look like DES all over again, even as to the reluctance of the government to stop the travesty, recalcitrance in the name of letting individual doctors decide what is best for their patients, no matter the science, no matter the ethics. If I did this work right—if I focused on truth and justice and not on my own misery—eventually I would help the CAH- and intersex-affected families and the ethical medical researchers, as I had originally set out to do.

So I crawled out of my hole and pulled myself forward. I started by carefully looking into the characters in the drama. I put out a series of feelers to find new leads. I tracked medical conferences and the medical literature to find any new data. I corresponded with strangers who sent me information, among them several mothers who wrote to tell me what they had been through with Maria New. (Such stories were only "anecdotal evidence," but they gave me leads.) I learned how to use the Freedom of Information Act (FOIA) to get copies of New's grants and information on the federal investigations we had initiated. When the OHRP FOIA office wouldn't give me what I asked for and the FDA FOIA office wouldn't even acknowledge my written requests, I hired a lawyer and sued the government. Meanwhile, when I had nothing new on dex on my plate to analyze, I kept up my reputation as a researcher and health and science writer by working on side projects. In private, I distracted myself with pro bono, confidential, medical-history-recovery work for individuals who had been medically traumatized, work that made me think I had some use to the universe, even if it ultimately turned out that I had been all wrong about dex. And day by slow day, week by slow week, I pressed on with investigatory historical research into dex.

After a few years of steady work, I could finally see clearly what had happened with dex. I could finally see where we'd been right, where we'd been wrong, and where we'd perhaps been played.

AS WE HAD FEARED, prenatal dex did turn out to be a story of how layers upon layers of medical research protection systems can fail—and fail even around the group recognized to be the most vulnerable: human fetuses. As we hadn't fully understood, prenatal dex also turned out to be a story of what results when you get a perfect storm of beneficent intentions, professional loyalties, gut-level fears of sexual anomalies, unmanaged conflicts of interest, and mindless bureaucratic paper-pushing. To my horror, prenatal dex also turned out to be something of an ethics canary in the modern medical mine; whereas I had assumed at the outset of my investigations that prenatal dex would always be a story of exception, over the years I learned that in fact the same story of failed protections—and failure of truly informed consent—may well be playing out in clinics all over the United States.

That's the short version of what I discovered in those three years of work. Here's the slightly longer one:

Maria New's main NIH grant, a grant to study many forms of and treatments for various adrenal disorders, had been consistently funded all the way back to the 1970s. When the idea of using prenatal dexamethasone to prevent virilization in CAH-affected females came along in the mid-1980s, New successfully rolled the intervention into her multiarmed grant renewal applications, thereafter using prenatal dex to give the NIH more reason to fund her work. She started offering prenatal dexamethasone for CAH through her Cornell clinic in 1986, and right around that time, presumably because she planned to publish on the intervention, she sought and obtained Cornell research ethics committee (IRB) approval to study what happens when you dose pregnant women and their fetuses with dexamethasone starting very early in pregnancy to try to prevent intersex in the female offspring. These pregnancy drug trials were not rigorous, in spite of the fact that the goal was to have the drug cross the placenta and change fetal development, in spite of the fact that seven out of eight of the offspring exposed in the first few weeks of embryonic development could stand no benefit and would assume all the risks. There was no placebo control, and no blinded assessments of results. In the mid-1990s, Maria New's team published some results, comparing girls who had been exposed to prenatal dexamethasone to sisters who had not; they claimed that this showed the intervention worked.

Contrary to what I had thought when we sent our complaints to the feds, New *did* seem to have consistently gotten IRB approval at Cornell for studies of prenatal dexamethasone for intersex prevention. However, New's IRB applications at Cornell, which I obtained via FOIA, throw up multiple red flags, flags which apparently did not attract the notice of the IRB members. For example, in 1985, in New's first Cornell IRB application to include (among many other endocrinological studies) a plan to study the use of prenatal dexamethasone for CAH starting at two to four weeks of gestation, New did not check the "special populations" boxes on the front page of the form to indicate that her subject population would include pregnant women and fetuses. Checking those important header boxes would normally have set off special layers of review. Checking those boxes would have ensured that absolutely everyone in charge of managing research subject protections knew that, deeper in the packet, there was a plan for fetal experimentation—and yet as New renewed her IRB approvals each year, no one serving on the Cornell IRB seems to have noticed these "special groups" boxes were not being checked. It took *sixteen years* for someone to notice.

The consent forms that the Cornell IRB approved for New's prenatal dex work also minimized the risks of this experiment in shocking ways. Any woman who did sign them could not legitimately be said to have given informed consent to put herself and her offspring into this experiment. Instead of describing the risks as essentially unknown, the 1985 consent form approved by Cornell described potential harms of dexamethasone exposure as "transient and reversible suppression of the maternal and fetal adrenal gland." A pregnant woman reading this description would have reasonably assumed any harm from prenatal dex would be—well, transient and reversible, especially given how the form went on to emphasize the relatively low dose and that "pregnant women and fetuses treated to date with this regimen have not experienced complications." There is no way, given the high risks, the unknowns, and the extremely controversial status of this intervention, Cornell should ever have allowed such a reassuring description as put forth in this form. But problems with the consent forms persisted year after year, including in that they did not contain appropriately updated information about findings of possible harm to mothers and children from prenatal dex. As late as 2004, Cornell's IRB was still approving consent forms that reassured pregnant women that prenatal dexamethasone "has been shown to be safe for the fetus" even while the supposed point of the study was to determine "long-term safety of prenatal treatment."

New repeatedly boasted to the NIH that a strength of her research team was that her clinic had far more CAH-affected and dex-exposed patients to use as research subjects than any other researcher. By her own admission, she had this advantage of numbers specifically because offering prenatal diagnosis and dex treatment brought CAH-affected families to her door. In her 2001 grant renewal application, for example, she told the NIH, "The prenatal diagnosis and treatment program has *provided a new source of patients* with CAH, adding about 15 patients per month. Clearly we are in a position to expand the [clinic's] database to provide sufficient patients to answer the [medical research] queries relevant to CAH." That particular NIH application included a table describing human subjects under the research arm called "prenatal dx [i.e., diagnosis] and treatment in families at risk." The total number of her research subjects for that study aim is stated as 2,144. So prenatal dex wasn't just a boon in and of itself in terms of clinic and research income; prenatal dex allowed New a big boost in numbers of CAH-affected people who could then be used to obtain research funding for studies beyond prenatal dex.

To get and maintain her NIH funding for experimentation involving exposing fetuses to dexamethasone, New had to show that she had Cornell's IRB's approval for such studies. As noted above, it appears that, indeed, she consistently got Cornell's IRB's approval for studies that would have involved pregnant women, and that she then reported to the NIH the IRB approvals. Keep in mind she didn't have to show the NIH signed consent forms from the mothers she was naming as subjects, or even the blank consent forms—just IRB approval to enroll subjects. The NIH would have assumed that the human subjects she was talking about in her grants had been properly enrolled—that they had signed well-vetted forms that meant they had been fully and honestly informed about potential benefits and risks before agreeing to become experimental subjects.

For many years, New's work chugged along, approved and funded annually. By 1996, the NIH was specifically funding New to study whether prenatal dexamethasone for CAH could "succeed" in making CAH-affected females, in effect, more likely to be straight and interested in becoming mothers. No one at the NIH seems to have raised a question about whether this was a reasonable goal of clinical care or medical research.

In August 2001, the young, Greek pediatric endocrinologist Kyriakie Sarafoglou was hired to work in the pediatric research program at Cornell's medical school. Under a new national program designed to provide on-site monitors for big NIH study locations, Sarafoglou was employed on NIH money as a "research subject advocate," overseeing patient safety at Cornell's Children's Clinical Research Center. Only a few months into that job, Sarafoglou realized she was seeing all sorts of problems with the $23 million NIH grant for which New was the principal investigator, including financial and ethical irregularities. For example, Sarafoglou noted concerning irregularities on New's IRB materials. Sarafoglou first tried raising the alarm within Cornell. This led to an internal investigation that found nothing especially concerning. The internal Cornell report, which I obtained via FOIA, praised Maria New for her storied career and suggested *Sarafoglou* needed replacing.

Unsatisfied with Cornell's response, Sarafoglou contacted the OHRP with her concerns, documenting them with thousands of pages of internal paperwork. Although OHRP proceeded to investigate, the agency did so at the usual glacial pace, perhaps because Sarafoglou had sent huge quantities of evidence of systemic problems in Cornell's pediatric research. With the investigation ultimately involving the Department of Justice, things got hot, and Sarafoglou was given notice that her position was being terminated. About a year after she had called on OHRP to investigate, Sarafoglou sent an anguished message to the agency: "It seems that my academic career is over. I just wish I had [had] the foresight to know the pain to me and damage to my career that filing with OHRP would cause."

Based on what I found when I FOIA'ed the OHRP investigation Sarafoglou initiated in 2003, it appears a major point of concern was IRB records at Cornell that made no sense. External reviewers looking at Cornell's pediatric research records found misleading consent forms as well as "at least one protocol in which a significant number of subjects were enrolled an [sic] on whom no documentation of informed consent could be found." One reviewer concluded that, although the people at Cornell clearly cared about children, "The IRB members interviewed did not appear to have a thorough understanding of the additional protections for children as subjects of research." In an angry memo that seems to have been written by Sarafoglou, someone complained that New was claiming "130 or 160 patients have been accrued each year [to New's IRB-approved CAH studies] but [there is] not one single dropout or patient studied under another protocol," a virtual impossibility. Yet the records from Cornell show New reporting to her IRB, in her annual applications for renewal, that not a single potential subject refused to be in, withdrew from, or had a complaint about her IRB-approved study protocol. Moreover, in a letter to her IRB, New claimed with regard to a study on prenatal dex for CAH that, "To our knowledge, there have been no adverse events in patients due to studies performed," even while she cited in that letter a publication she had co-authored which had reported adverse events for exposed mothers and their offspring! Incidentally, all medical researchers know that IRB-reportable "adverse events" include *any* medical problems while in the study, not only those you can pin to the intervention. So what are the odds studies involving subjects who are pregnant women and developing fetuses would not involve *any* reportable adverse events? Again, this is virtually impossible.

From 2003 to 2005, in response to Sarafoglou's whistleblowing, OHRP ultimately undertook a massive review of Cornell pediatric research. OHRP found so many problems, they took the extraordinary step of requiring review by Cornell's IRB of all active pediatric research at Cornell. One of the projects about which OHRP had pointed questions at that time was New's supposed research on prenatal dexamethasone for CAH. In a letter dated May 24, 2004, OHRP asked Cornell to "Please clarify whether the subjects described in the *Journal of Clinical Endocrinology and Metabolism* [1995 publication] entitled 'Extensive Personal Experience: Prenatal Treatment and Diagnosis of Congenital Adrenal Hyperplasia[']. . . were enrolled in a research study. If they were, please provide a copy of the complete IRB-approved protocol for these studies." What OHRP was asking was not just whether Cornell's IRB had given New *permission* to do a study, but whether she had actually had the hundreds of pregnant women research subjects *enrolled* in a formal study by signing the consent forms. After a lot of back-and-forth in which Cornell said they were having trouble finding the records, and in which Cornell reminded OHRP that New was now gone from Cornell, OHRP just decided to stop tracking what had happened to the pregnant women given dex. They just had too much else to sort out.

This is troubling—OHRP dropping the question of the ethics of the matter. But what is more troubling is that when, in response to OHRP's inquiries, a Cornell administrator asked New about the pregnant women mentioned in that publication, New wrote back a curt memo informing the Cornell IRB rep that, "Prenatal diagnosis and treatment is a standard procedure at the patient's request and has been performed in France since 1981 and in our lab since 1986." In other words, she was stating that so far as she was concerned prenatal dex wasn't experimental—so it didn't require consent to research. It was "standard" practice given because patients "requested" it.

This, I think, may explain why I can't find anything, in all the records, indicating actual enrollment numbers for pregnant women in IRB-approved trials of prenatal dex. Although she got IRB approval for fetal research as required to obtain NIH funding, in practice, New seems to have treated dex as "a standard procedure [performed] at the patient's request." Although OHRP would later tell us that "written informed consent (which included disclosure of risks) was obtained from subjects for participations in these studies," I can't find any consent forms that appropriately disclosed risks, and I can't find any direct confirmation of enrollment of pregnant women into consented research for intersex prevention. All we have is Cornell's word to OHRP that they looked into it and nothing had gone wrong. Recall that Cornell's internal investigation in 2002 also said nothing had gone wrong, and when OHRP came and did a real site investigation, they found much evidence to the contrary.

If Cornell's IRB administrators had audited their pediatric researchers' records prior to Sarafoglou's complaints, they might have picked up that the pregnant women being named in New's publications might not have given appropriately informed consent to research. But it appears that no such auditing occurred at Cornell's Children's Clinical Research Center (CCRC), a unit for which New herself served as director. Via FOIA, I have obtained extensive records and correspondence on IRB review of pediatric research at Cornell during this period, and I can find no reference to any audits—the kinds of inquiries that would have checked to see if people were actually signing consent forms—until Sarafoglou's complaints. The extensive OHRP correspondence resulting from the 2003-2005 investigation also makes no mention of regular internal audits, as one would expect if such audits had occurred. And the previously mentioned angry whistleblowing memo specifically decried the complete absence of regular audits, seeming to confirm that no pediatric research at Cornell was being audited during this period to ensure appropriate consent practices:

> [New's IRB protocol] like all the protocols performed in the CCRC has never been through a SAC [scientific advisory committee] review or an IRB audit . . . The Dean's Office and IRB may indicate to regulatory agencies that Cornell has made "numerous corrective actions" or are "developing new procedures for the IRB" or "initiating an audit program" . . . But these are EMPTY WORDS that do not reflect the reality at Cornell. If Cornell really bothered to perform all the things they say they will when denying charges to regulatory agencies, don't you think they would have at least properly audited a few protocols in the CHILDREN'S clinical research center, especially protocols begun before 1998.

Whatever she told the mothers, New consistently led the NIH to believe prenatal dex was experimental. Odds are NIH wouldn't have funded it if she'd told them it was standard of care. The portrayal of dex as experimental for grant-getting purposes didn't change when, shortly after leaving Cornell, Maria New landed herself a position as a pediatric endocrinologist at Mount Sinai School of Medicine, across New

York City. In 2006, in a research progress report to the NIH from her position at Mount Sinai, New assured the NIH that, even though she had left Cornell, her research on fetuses continued unabated: "The prenatal diagnosis and treatment program has galloped ahead so that we have now diagnosed and[/]or treated 768 fetuses." She specifically referred to fetuses being "treated" with prenatal dex at Mount Sinai as part of her *research* progress. Because NIH funding requires proof of IRB approval for grant-funded experiments, New got a letter from the head of Mount Sinai's IRB to give to the NIH, indicating Mount Sinai's IRB had approved the study of "prenatal diagnosis and treatment." This letter was signed by the same administrator who told the OHRP in response to our complaints that there was no fetal experimentation of this kind occurring at Mount Sinai.

It is possible that what happened at Mount Sinai is what I think happened at Cornell: New obtained IRB approval for a study of prenatal dex to show to the NIH, as the NIH would have required to fund her prenatal dex studies, but pregnant women given dex for CAH virilization prevention were, at Mount Sinai, probably treated like patients, not asked to sign consent to fetal research. This seems to be confirmed by what that Mount Sinai administrator told OHRP in response to our complaints: "the decision as to whether to treat prenatally with dexamethasone or not was made by the patient and her physician," not under the auspices of an IRB-approved trial. So even though New very specifically represented to the NIH that these pregnant women and fetuses were human subjects of her research at Mount Sinai, they were probably afforded no protection by an IRB, because they were "just" patients when they were offered dex.

Some might wonder, how did New even manage to get hired at Mount Sinai after the NIH and OHRP investigations of her work at Cornell? Just after she had to relinquish her leadership positions and income at Cornell, New had written to the NIH asking for money to support herself under the NIH Merit Award System. New pleaded to her friends at the NIH that, "As I am no longer Chairman of Pediatrics,

Chief of Pediatric Endocrinology, and Program Director of the CCRC, I have much more time to devote to the research," and added, "This is a hard time for me and I am deeply appreciative of your consideration." Noting that "her circumstances changed abruptly this year when she had to relinquish" her leadership positions "and the salaries entailed in these positions," her colleagues on the NIH staff "enthusiastically support[ed]" giving her the award and writing her a big check. The NIH kept funding her. With a big active NIH grant, and with a brand new Merit Award to her name, Mount Sinai probably saw Maria New as the goose that would keep laying golden eggs. The NIH-Cornell fraud settlement was only announced after New was safely ensconced at Mount Sinai, and by then, OHRP had dropped its questions about her research ethics. So, without any official blemish on her name, she just moved over to Mount Sinai and "galloped ahead" on prenatal dex. Continuing to advertise the intervention as "having been found safe for mother and child," she drew in more and more families to her relocated clinic. Because she had so many more CAH-affected research subjects than anyone else, the NIH kept funding her. According to one medical school administrator who looked at what I found (one to whom I happen to be married), Maria New seems to have invented a perpetual motion machine of NIH funding.

I think it is fair to summarize what happened this way: *The pregnant women given dexamethasone under Maria New's guidance may never have understood they were in a high-risk experiment, and the* NIH *may never have understood they weren't.* Most of the pregnant women, like those who told their stories to the *Time* reporter, probably thought they were taking a drug that was safe and effective, not experimental. And the NIH, being shown by New the official IRB approval for her studies and being told about the "progress" of fetal treatment research, probably assumed that all the women were being actively enrolled in careful experiments that would be meticulously overseen by the medical schools' IRBs. Because the medical schools' IRBs appear never to have audited New's practices—at least until the investigations at

Cornell—they do not seem to have been keeping careful track of what was going on.

It is quite possible that the problem of blurring the line between clinical care and human experimentation in Cornell's pediatric unit went beyond Maria New. Internal minutes from Cornell's IRB show that, after Sarafoglou raised the alarm, a member of the IRB who had simply been signing off on New's annual renewals said this: "one of the issues the Program Review Committee is reviewing is the blurring of patient care vs. research and whether patients are being told which procedures are part of their care and which are part of the research."

THAT'S WHAT I NOW BELIEVE happened with New's work on prenatal dex. Given that I was able to figure all this out—given that I learned all this chiefly through obtaining government records through the Freedom of Information Act (FOIA)—why didn't the government do something when we made our complaint in 2010? That is a tale in and of itself.

Recall that it was late 2009 when my allies called me to help. In February 2010—not realizing there had ever been an OHRP investigation into New's work on prenatal dex—Ellen Feder, I, and thirty of our colleagues appealed to the FDA and OHRP.

About three months later, *AJOB* released the target-article manuscript by Larry McCullough and Frank Chervenak attacking us as "unethical" and "transgressive" for our calls for a federal investigation. FOIA confirmed that, as I had suspected, Larry McCullough had sent the *AJOB* manuscript to OHRP while that agency was deciding what to do with our complaints. In his cover letter, McCullough told OHRP, "We hope that your [*sic*] will take our critique into account in your deliberations." None of the authors' conflicts of interest were disclosed to OHRP. For the target article, McCullough listed Baylor College of Medicine as his only affiliation. As it turns out, he also had two other active, paid faculty positions: at the medical schools of Cornell and Mount Sinai. The Cornell chair of Obstetrics, Frank Chervenak, coauthor of the target article, did not mention to OHRP that he had served as "key personnel" on Maria New's NIH grant.

After receiving our letters of concern, the FDA gave the job of the prenatal dex investigation to a physician-ethicist named Robert "Skip" Nelson. As I figured out much too late to do anything about it, not only was Nelson serving on the *AJOB* editorial board (with Larry McCullough), but he was also actively negotiating a new *AJOB* journal editorship in chief for himself, for a new offshoot journal called *AJOB-Primary Research*. Just to be clear, he was in negotiations with *AJOB* *while* he was running the federal investigation that *AJOB* was being used to undermine. Nelson's new editorship came with a fancy title and ten thousand dollars a year, to hire an editorial assistant. In the e-mail exchange in which he informed his ethics supervisors that he was planning to accept this position with *AJOB*, Nelson said he hoped it would allow him to get an adjunct faculty appointment at Georgetown University's ethics unit, but never mentioned what role the journal was playing in the FDA investigation he had been given to run. Remember the "coincidence" of *AJOB* officially publishing the McCullough et al. target article just as the federal findings were made public? E-mails discovered via FOIA suggest that Nelson, by then working for both FDA and *AJOB*, was keeping track of when the federal findings would be released.

Nelson's behaviors would be less concerning if what he had represented in his official findings had been factually accurate. In one of the most important revelations of the whole federal investigation, in the name of the FDA, Nelson had told everybody that Maria New sought and obtained an IND (investigational new drug) exemption from the FDA in 1996 for a study using prenatal dexamethasone to prevent virilization in female fetuses. This would have meant that way back in 1996, a high-ranking physician at the FDA had reviewed the intervention and had formally decided that it was not risky enough to require significant FDA review before proceeding.

This seemed almost impossible—the FDA giving a pass on an experiment *meant* to try to change fetal development, with seven of eight fetuses deriving no possible benefit? When Nelson made the claim in 2010, we all believed it. When I obtained a copy of the 1996 exemption letter, however, I found that it actually refers to a very different thing. It refers to a proposal from Maria New "to utilize dexamethasone to treat pregnant women with a [*sic*] congenital adrenal hyperplasia." Based on this wording, it looks quite possible that what New was proposing to the FDA in 1996 was to use dexamethasone on pregnant women who *themselves* had CAH; women with CAH often need to have their disease treated during pregnancy, to maintain their own health, and dexamethasone can be used to treat the disease. In other words, the exemption was probably not, as Nelson claimed, for a study of "the administration of dexamethasone during pregnancy *for the purpose of preventing virilization* in females with congenital adrenal hyperplasia." It appears to have been meant for a study of using dexamethasone during pregnancy *for the purposes of maintaining the health of pregnant women who themselves had CAH*. That's presumably why the physician at the FDA didn't flip out at what New had proposed to do without full FDA review. What was most likely being proposed wasn't a fetal engineering experiment!

When I contacted the physician who had signed the letter back in 1996, he had no recollection of the matter. But for his part, in 2010, Nelson knew full well there was no documentation to support his public misrepresentation of the letter. In internal communications, Nelson told OHRP staff that the single vague FDA letter was the only record of all this, so they could know no more than what the letter said. He explained to OHRP that any related documentation (which would have explained exactly what New had proposed) had been shredded during an FDA move. But in his 2010 memo to OHRP and the public, rather than providing what the IND exemption letter actually said—that New had been given an exemption to study "dexamethasone to treat pregnant women with a [*sic*] congenital adrenal hyperplasia"—he made it look as though the FDA had already reviewed prenatal dexamethasone for virilization prevention in 1996 and had found it to be nothing especially concerning.

If OHRP had done a full investigation in 2010—if the agency had bothered to look up its own 2003 investigation at Cornell, as I later did—I think it would have found as much reason to be concerned as we had. Instead, an e-mail obtained via FOIA of the 2010 investigation shows the head of OHRP thanking Nelson for suggesting that OHRP rely on his work. That is what OHRP did—relied largely on Nelson for OHRP's conclusions.

OHRP was not alone in heavily relying on Nelson's review. A couple of months after the federal findings emerged, *AJOB* published a piece by Maria New extensively quoting Nelson's FDA memo and concluding, "The recent reports by the Office of [*sic*] Human Research Protections and the FDA therefore make crystal clear that my research on prenatal treatment of CAH is and always has been both legally and ethically proper at every level." The article included no disclosure of the FDA official's (Nelson's) relationship with *AJOB*. When we asked *AJOB* to make this disclosure and to correct New's article's title to make clear that the federal government had not "vindicated" the intervention, the editors refused. Thanks to Nelson and the rest of the *AJOB* editorial team, the medical literature continues to include a supposedly peer-reviewed article that says that the FDA has vindicated prenatal dexamethasone for CAH.

When we set out to petition the FDA, I felt sure that the agency would at least stop Maria New from advertising prenatal dexamethasone for CAH as having "been found safe for mother and child," because the language seemed to be the kind you're only allowed to use for FDA-approved drug indications. New is subject to no such limitation. Thanks to a loophole New is still driving a truck through, only two classes of people—employees of a drug's maker and researchers with active FDA investigator status—are prohibited from advertising an off-label use as "safe and effective." Because New doesn't have appropriate FDA investigator status for this intervention, she isn't subject to the prohibition. Under current regulations, if you unethically study a drug use, you can also unethically advertise it. As a Hail Mary play, I tried appealing to the "bad-ad" division of the FDA. The staff of that division never even acknowledged my letter.

SWEDEN REMAINS THE only place where prenatal dexamethasone for CAH has been studied using a prospective, long-term tracking approach with full ethics board oversight and informed consent throughout. Just two months after the U.S. agencies decided our letters of concern were unfounded, the Swedish clinical researchers studying prenatal dex were so alarmed by their results that they went to their ethics committee to tell them they would no longer provide the intervention in Sweden even if a woman was willing to sign up for a prospective clinical trial. They shut it down.

The Swedish group continues to track those already dosed. In a study of forty-three children given prenatal dex for CAH, results show that in the quest to sex-normalize the females, boys who are "accidentally" exposed may be genitally and behaviorally feminized. This study also confirmed earlier findings of impaired verbal working memory in exposed children who turned out not to have CAH. In terms of major side effects among the forty-three children tracked, researchers documented a total of eight "severe adverse events." This is an astonishingly high rate, albeit one that requires a larger study to establish whether these problems were caused by the drug. The "events" included growth disorders like the one described by the American mother in the *Time* magazine article, a mood disorder requiring hospitalization, and three cases of mental retardation. (Just to be clear: 7 *percent* of the prenatally exposed children in this prospective cohort have mental retardation. That's about ten times the normal rate.) In response to these findings, the Swedish team has declared it "unacceptable that, globally, fetuses at risk for CAH are still treated prenatally with DEX without follow-up."

In 2012, two years after we made our complaints to the feds, New's research group, led by Heino Meyer-Bahlburg, published a retrospective convenience-sample study of prenatal dexamethasone for CAH. The group looked at cognitive function in sixty-seven children prenatally exposed, including eight CAH-affected girls and fifty-nine boys and CAH-unaffected girls, a tiny portion of the number of children New told the NIH she has had in her prenatal treatment studies. While the new study's finding "contributes to concerns about potentially adverse cognitive after effects of such exposure," in sharp contrast to the Swedish data, this study by New's group found some *positive* outcomes in terms of brain function—i.e., the exposed children appeared to do *better* on some measures than children never exposed! The paper noted that the result was confusing, and considered what it might mean. Among those considerations was not the most obvious possibility: the sample was highly skewed.

FOIA provided us a copy of the design for the retrospective follow-up study to be carried out with Heino Meyer-Bahlburg. This is how I discovered that the design specifically stated as one of the "exclusion criteria for all groups" this: "mental impairment which prevents understanding of questionnaire." In Sweden, 7 percent of those exposed show mental retardation. In the United States, those who may be significantly cognitively damaged by the drug won't even make it into the study.

In the United States, pregnant women offered prenatal dexamethasone for CAH remain essentially unprotected by the Office for Human Research Protections and the Food and Drug Administration. Doctors who believe the intervention is safe and effective can even advertise it as such. (Maria New still does.) If an American mother or her child is harmed by it, the family's only recourse is to sue. In order to win such a lawsuit, the family would need to prove that the harm was caused by the prenatal dexamethasone. Without high-quality scientific studies of the intervention—prospective, long-term studies involving large enough numbers of subjects to show statistical significance—this would be almost impossible.

Back in September 2010, when I was acting like a beaten dog in response to the federal findings—before I took another three years to figure out what really happened—my dex-worried friend David Sandberg at the University of Michigan said to me something very useful: The people we've always really cared about are the mothers and the children, not Maria New. We can't stop her, but maybe we can make sure everyone knows the truth. David asked me to keep working, specifically to see if I could make sure there wasn't a single obstetrician, genetic counselor, or pregnant woman considering prenatal dexamethasone for CAH who didn't know that this drug use is experimental and dangerous.

This is not how informed consent is supposed to work, but as I write, if you Google "prenatal dexamethasone for CAH," that's the message you get. The top hit is a scholarly article I coauthored with Ellen Feder and Anne Tamar-Mattis documenting what really happened in the history of this intervention. The second hit is a support group page discussing the "controversy." The third hit is a report I co-wrote for the medical news literature letting doctors know that prenatal dex continues to be considered experimental and in need of IRB oversight. The fourth is the *Time* magazine article. The fifth is the Swedish report of harm. The sixth is the *Bioethics Forum* article in which we documented New's use of prenatal dex to try to prevent lesbianism and tomboyism.

The seventh is an article I wrote for *The Atlantic*. It's about how I discovered by accident, from a pregnant woman writing to me, that fertility doctors are using dexamethasone on their pregnant IVF (in vitro fertilization) patients on the hunch that it might prevent miscarriages. As was the case with diethylstilbestrol (DES), there is no evidence that dexamethasone is effective or safe for miscarriage prevention. There are no studies of this use. I tried to find an investigative health reporter willing to research and write this story. Finding none, I did it myself. My editor at *The Atlantic* had previously told me—while we were having coffee literally in the shadow of the Watergate Hotel—that I should avoid investigative work. He said *The Atlantic*'s Health section simply didn't have the resources to support it. Perhaps because he has an MD, he published this one anyway.

What do you do when you realize the Fourth Estate is dying, and there are no accountants on white horses in Washington? The answer now seems obvious to me: You work harder. You find compatriots in the ivory tower—the kind of people who don't mind difficult questions that may take years to answer and that might get you in a heap of trouble—and you work, together, as long and as hard as you can.

CONCLUSION

TRUTH, JUSTICE, AND THE AMERICAN WAY

IN THE LATE 1980s, the Polish labor movement known as Solidarity grew so powerful that it did what many had thought impossible: It peacefully forced the Iron Curtain back. In my own family, this meant not only a blessed end to the occupation that had lasted since the Allies had handed Poland over to the Soviets to keep the peace; it also meant, more practically, that my mother could go back to Poland for the first time since she had left, in 1947, at the age of eleven.

In the summer of 1990, my older brother, my father, and I went with her. I was twenty-four, two months shy of moving to Indiana to start my new life in the academy. This was an astonishing time to be in Poland. A nation that had long been silenced and used by the Soviets had burst forth suddenly into noisy liberation. A country that had been painted gray by years of corrupt socialism was rapidly painting itself red and white again.

We found ourselves one day in a crowded city square—I think it was in Gdansk, a city on the Baltic coast—where an enormous crowd had gathered in happy anticipation. A ritual that was occurring all over Poland was about to be enacted here. A large crane had been used to carefully hoist a big statue of Lenin that for decades had been watching over the square. Ostensibly this statue was in the process of being moved by workers to a "safe location," but everyone knew what was really going to happen. At the appointed time, the crowd drew back, the crane operator raised the statue a little higher, and the giggling crowd muttered the Polish equivalent of oopsie-daisy. A moment later, the crane operator raised his arm ceremoniously and pushed a button, dropping the statue and smashing it into a thousand pieces. A roar went up, bottles came out of jackets and bags, and singing commenced. Freedom. Were it not for the big Polish mustaches and all the vodka, I would have thought I was in Philadelphia in July 1776. In any case, I suddenly felt lucky to have grown up always conscious of the stark contrast between Poland under the Soviets and my life in America.

A couple of days later, we spent some time visiting my mother's extended family in the place where she had lived in poverty throughout the war, a tiny farming village still comprised of only a handful of families. I was trying to follow the rapid-fire, happy conversations, but having ever learned only a few words of Polish, I spent most of my time trying to remember how I was related to each of these people. In the 1980s, some of the older ones had managed to visit us on tourist visas. As I watched these great aunts and uncles now sharing food with my parents, I remembered these same people sitting in our Long Island kitchen as my mother sewed twenties into the linings of their coats; she had explained to me that American currency would go a long way on the black market when they got home.

Noticing my brother and me sitting so quietly, my mother suggested that we take a walk with her around the area. (My father's multiple sclerosis made it impossible for him to walk more than a few feet, so we left him with our cousins.) On the walk, our mother kept exclaiming how small the hills appeared to be compared to her childhood memories. I reminded her jokingly that there's a reason for Poland's name. (It means "land of the plains.") My brother joked back that he was under the impression Poland's name actually translated to "invade here."

Suddenly, on a little hill my mother seemed to specifically remember, we came upon a perfectly tended single grave. I couldn't make sense of this. First of all, Poles bury their dead in proper cemeteries, not on random hills. Second, it was pretty clearly the grave of a *German* soldier from the war, and yet someone had been carefully keeping it up all those years; the surrounding grass had been cut back, and flowers were placed around the stone.

I asked my mother what all this meant. She told us that, during the war, a village nearby had been found to be harboring Jews, and a Nazi officer had given one of his soldiers an order to kill everyone in the village. This was standard practice, meant as punishment and as a warning to others. But the soldier had refused. As a result, his officer had shot him dead, on the spot. Those who had somehow survived had honored the soldier with this burial.

My mother then quietly mumbled something religious in Polish and made the sign of the cross. Involuntarily, I made the same gesture.

When I was about sixteen, right around the time I gave up on Catholicism, my mother gave me a rare glimpse into the worldview presumably bequeathed to her by this war. While we were folding laundry together, perhaps knowing the guilt I was feeling over losing my religion, she suddenly said to me that she didn't think dogma was particularly useful in most instances. In most situations in which you find yourself, she said, if you could stop and think clearly—think beyond your immediate self and beyond your predictable loyalties—you could tell what was the right thing to do and what was the wrong thing. Being good, she observed, meant being good to others, including strangers. And that was pretty much enough to live by. But how can you know the right thing to do? Human reasoning, she said—referring now explicitly to Socrates and Plato—human reasoning is imperfect. Human bias keeps us from perfect vision of what is happening around us. But the quest for truth—the quest to understand the world around us—must ultimately be how you enact the good.

. . .

WHEN I JOINED the early intersex-rights movement, although identity activists inside and outside academia were a dime a dozen, it was pretty uncommon to run into *evidence-based* activists. Even rarer were data-oriented scholars who purposely took on advocacy work. Today, all over the place, one finds activist groups collecting and understanding data, whether they're working on climate change or the rights of patients, voters, the poor, LGBT people, or the wrongly imprisoned. It's also pretty easy to find university-based, data-oriented scholars in medicine, climate studies, and law who spend much of their time out of the ivory tower going to legislatures, courts, and policy meetings to advocate for marginalized individuals and endangered populations. Attention to evidence in the service of the common good is at perhaps an all-time high. People doing advocacy are smarter, and the smartest people often now do advocacy.

The bad news is that today advocacy and scholarship both face serious threats. As for social activism, while the Internet has made it cheaper and easier than ever to organize and agitate, it also produces distraction and false senses of success. People tweet, blog, post messages on walls, and sign online petitions, thinking somehow that noise is change. Meanwhile, the people in power just wait it out, knowing that the attention deficit caused by Internet overload will mean the mob will move on to the next house tomorrow, sure as the sun comes up in the morning. And the economic collapse of the investigative press caused by that noisy Internet means no one on the outside will follow through to sort it out, to tell us what is real and what is illusory. The press is no longer around to investigate, spread stories beyond the already aware, and put pressure on those in power to do the right thing.

The threats to scholars, meanwhile, are enormous and growing. Today over half of American university faculty are in non-tenure-track jobs. (Most have not consciously chosen to live without tenure, as I have.)

Not only are these people easy to get rid of if they make trouble, but they are also typically loaded with enough teaching and committee work to make original scholarship almost impossible. Even for the tenure-track faculty, in the last twenty years, universities have shifted firmly toward a corporate model in which faculty are treated as salespeople on commission. "Publish or perish" was the admonition when I was in graduate school, but today the rule is more like "external funding or expulsion." (I am, as I write, facing this myself at Northwestern.) Our usefulness is not measured by generation of high-quality knowledge but by the volume of grants added to the university economic machine. This means our work is skewed toward the politically safe or, worse, the industrially expedient. Meanwhile, administrators shamelessly talk about their universities' "brands," and lately some are even checking to see if their faculty are appropriately adhering to "the brand." Yet more evidence of a growing and scary corporate mentality. Add to this the often unfair Internet-based attacks on researchers who are perceived as promoting dangerous messages, and what you end up with is regression to the safe—a recipe for service of those already in power.

The pressure to get ever more grants and to publish early and often has also led to a system wherein work is often published before it's ready—before it is really finished and, more important, before it is really checked. Indeed, as I've wandered from discipline to discipline, I have again and again come across a stunningly lazy attitude toward precision and accuracy in many branches of academia. (Legal scholarship is the one notable exception.) Outsiders to academia would probably be shocked to overhear the conversations I have with science, medicine, and humanities scholars and journal editors about the need to fact-check work. It's not that they argue with me. *They ask me what I'm talking about.* The push is to get the work out, to get that publication line on your productivity report, to score the high impact factor—all goals born of a system that supports individual competition for funding over the common need for a reliable body of knowledge. Who needs fact-checking when accuracy is not rewarded and sloppiness is rarely punished?

Perhaps most troubling is the tendency within some branches of the humanities to portray scholarly quests to understand reality as quaint or naive, even colonialist and dangerous. Sure, I know: Objectivity is easily desired and impossible to perfectly achieve, and some forms of scholarship will feed oppression, but to treat those who seek a more objective understanding of a problem as fools or de facto criminals is to betray the very idea of an academy of learners. When I run into such academics—people who will ignore and, if necessary, outright reject any fact that might challenge their ideology, who declare scientific methodologies "just another way of knowing"—I feel this crazy desire to institute a purge. It smells like fungal rot in the hoof of a plow horse we can't afford to lose. Call me ideological for wanting us all to share a belief in the importance of seeking reliable, verifiable knowledge, but surely that is supposed to be the common value of the learned.

What privilege such people enjoy who can say there is no objective reality, no way to ascertain more accurate knowledge! I know from experience, these are people who typically claim to speak on behalf of the marginalized and the oppressed, yet they have not sat and learned enough anatomy and medicine to know what a clitoroplasty actually involves in terms of loss of tissue and sensation; they have not witnessed what happens to the minds and bodies of scholars falsely accused of crimes against humanity; they have not had to watch prenatal dexamethasone being advertised as "safe for mother and child" while knowing from a literature search that there is not a single decent scientific study to support such a claim. If they don't want to believe that there's an objective reality as to what a glucocorticoid does to a four-week-old embryo, then how are they to understand how much the truth matters to justice? These must be people who have never had to fear enough to desperately need truth, the longing for truth, the gift of truth. Surely, the "scholar" who thinks truth is for children at Christmastime is the person who has never had to fear the knock of the secret police at her door.

. . .

GALILEO COULD NOT have been put under house arrest for suspicion of grave heresy in America. Sounds banal, I know—but as we face a time when we are mature enough to understand that activism aimed at social progress but conducted without facts amounts to a pointless waste of resources, and the people best poised to do fact-collection are under siege from multiple sides—this is worth meditating on for a moment.

Galileo could not be arrested for suspicion of grave heresy in America. Why is that?

The reason is that the activists who founded the United States—the Founding Fathers—understood the critical connection between freedom of thought and freedom of person. They understood that justice (freedom of person) depends upon truth (freedom of thought), and that the quest for truth cannot occur in an unjust system. It's no coincidence that so many of the Founding Fathers were science geeks. These guys were rightly stoked about the idea that humans *working together* had it in their power to know and to improve the world—scientifically, technologically, economically, politically. These were men of the Enlightenment who had broken through dogma into a fantastic new vision for humankind: crowdsourcing. No longer would knowledge and power flow from top down, following archaic rules of authority and blood inheritance. In science as in political life, the light of many minds would be brought to bear to decide together what is right and is just. In such a system, a man arguing for a new vision of the universe could never be arrested merely *for the argument*, no matter how much it threatened those in power.

So here's one tiny historical story I wish that activists and scholars today would return to—because it would help them see why they have to value the same things.

America, 1776: The Declaration of Independence is finally signed, formally renouncing British rule and articulating a new vision of human liberty. There was no Internet in 1776, of course, to spread the good news, so the news had to be spread by other means. Knowing that the people of Philadelphia needed to be informed of the monumental turn of events, John Nixon climbed up on a platform and read the declaration aloud to an assembled crowd.

Here's the thing: The platform Nixon climbed up on had been erected seven years earlier, and not for political purposes. It was put up in 1769 by the American Philosophical Society (the scientific society founded by Benjamin Franklin) so that one of their members, the astronomer David Rittenhouse, could make formal observations of the transit of Venus—the passage of that planet across the face of the sun. Rittenhouse was working with the support of his peers to further the post-Galilean understanding of the universe. The platform was put up to look at the sky.

One scientist in 1769, following in Galileo's footsteps, wanting simply to know more about the stars, had literally set the stage for another man in 1776 speaking to the people the truth of their core freedoms.

If—as the investigative press collapses and no longer can function as an effective check on excess and corruption, and people live and die forever inhabiting self-obsessed corners of the Internet, and the government and the ad-selling Google industrial complex ever increase surveillance on us, and we can't trust people in the government to be our advocates or even to be sensible—if we have any hope of maintaining freedom of thought and freedom of person in the near and distant future, we have to remember what the Founding Fathers knew: That freedom of thought and freedom of person must be erected together. That truth and justice cannot exist one without the other. That when one is threatened, the other is harmed. That justice and thus morality *require* the empirical pursuit.

I WANT TO SAY TO ACTIVISTS: If you want justice, support the search for truth. *Engage* in searches for truth. If you really want meaningful progress and not just temporary self-righteousness, *carpe datum.* You can begin with principles, yes, but to pursue a principle effectively, you have to know if your route will lead to your destination. If you must criticize scholars whose work challenges yours, do so on the evidence, not by poisoning the land on which we all live.

To scholars I want to say more: Our fellow human beings can't afford to have us act like cattle in an industrial farming system. If we take seriously the importance of truth to justice and recognize the many forces now acting against the pursuit of knowledge—*if we really get why our role in democracy is like no other*—then we really ought to feel that we must do more to protect each other and the public from misinformation and disinformation. Doing so means taking on more responsibility to police ourselves and everybody else for accuracy and greater objectivity—taking on with renewed vigor the pursuit of accurate knowledge and putting ourselves second to that pursuit.

I know that a lot of people who met me along the way in this work thought I'd end up on one side of the war between activists and scholars. The deeper I went, however, the more obvious it became that the best activists and the best scholars actually long for the same kind of world—a free one.

Here's the one thing I now know for sure after this very long trip: Evidence really is an ethical issue, the most important ethical issue in a modern democracy. If you want justice, you must work for truth. And if you want to work for truth, you must do a little more than wish for justice.

EPILOGUE

I KNEW WE'D TURNED A CORNER on intersex rights when I attended a support-group meeting in 2010 and got to hear a high school girl tell the story of how she'd recently found out she was born with testes. This girl—a very pretty, Christian, cheerleader type—told of being taken by her mother to the doctor because, although she was obviously maturing, girl into woman, she had never menstruated. The doctor did a pelvic exam and soon realized the girl had an intersex condition—one that meant that, even though she had developed as a near-typical female in genital anatomy and brain, even though she was sexually maturing in appearance like most girls, she had a Y chromosome and testes internally (and no ovaries or uterus). She was born with a genetic condition that left her lacking the androgen receptors required for cells to respond to the masculinizing hormones being made by her body, so even though she had a Y chromosome, her body had developed close to the female-typical pathway except for some of her internal organs, those whose development depends on something other than androgens.

When the doctor realized what was going on with this girl's body, instead of withholding information from the patient—as was, until recently, pretty standard—the doctor told her the truth. In fact, not only did that doctor tell her the truth, the physician put her in touch with other young women with the same condition so they could offer her loving peer support. No shame, no veil of total secrecy. Before the gathered audience in which I sat, in which women like her probably outnumbered women like me, this teenager concluded by saying that not only was she not upset that she had been born with this condition, she was grateful for it. Being born this way had taught and given her so much.

Although this young woman's experience is not yet universal in America, telling the truth is becoming the treatment norm. This progress can be traced to the intersex patients' rights movement started by affected individuals like Bo Laurent, a movement directly inspired by the movements for women's rights, civil rights, and gay and lesbian rights. Physicians are finally understanding that when it comes to treating people born with intersex conditions, attributed shame and psychological isolation form the basis for *unnecessary* trauma. In 2010, the very year that young woman spoke of her awakening, a prominent American surgical specialist in intersex care openly acknowledged in the *Journal of Urology* that the real problem with intersex is not the child, but the way the child is treated in the clinic: "Secrecy is a recipe for shame, low self-esteem, and psychological disaster, and is to be avoided. . . . It may be that long-term psychological support is of equal or superior importance to the anatomical result [of surgery]."

In Europe, progress seems to be picking up even more quickly. In 2012, following successful political pressure from Swiss intersex activists, the Swiss National Advisory Commission on Biomedical Ethics issued a special report regarding "ethical issues related to 'intersexuality,'" recommending first and foremost that

> the suffering experienced by some people with DSD [differences of sex development] as a result of past practice should be acknowledged by society. The medical practice of the time was guided by sociocultural values which, from today's ethical viewpoint, are not compatible with fundamental human rights, specifically respect for physical and psychological integrity and the right to self-determination.

In an even more startling turn of events, in 2013, the United Nations special rapporteur on torture recommended that legal actions be taken to stop "forced genital-normalizing surgery" on children too young to consent. The special rapporteur, Juan E. Méndez, said he wished to shed "light on often undetected forms of abusive practices that occur under the auspices of health-care policies," including abusive "medical" treatments wrought on gender-nonconforming children, gay children, and intersex children.

In the United States and Europe, boys and girls born with sex anomalies are still routinely subject to risky genital-normalizing surgeries that are often medically unnecessary, unsupported by scientific evidence, and difficult or impossible to reverse. When challenged, many American specialists today say that it's *parents*, not clinicians, who insist on surgery. Nevertheless, a recent study backs up what I've seen in practice: When faced with decisions about genital-normalizing surgeries, parents are likely to go whichever way the clinician advising them is leaning, and in America, most clinicians still lean toward "corrective" surgery. In Europe, where the evidence-based medicine movement is powerful, in part because medicine is socialized, clinicians appear more inclined to a conservative, "first do no harm" approach. Unfortunately, I expect it will continue to be very difficult to get American clinicians who see these interventions as beneficent to pull back except through threat of legal action.

As I write, there is a case wending its way through the American courts brought on behalf of a young boy born with ambiguous genitalia who was surgically made to look female at the age of sixteen months. The surgery occurred at the order of the state of South Carolina while the baby was in foster care. Sometime after the surgery, the child was adopted by supportive parents, who accepted the child's eventual self-identification as a boy. The family's lawyers—including Anne Tamar-Mattis, who provided legal assistance to our efforts around prenatal dexamethasone—are arguing that the boy was denied due process as accorded by the Fourteenth Amendment when surgeons fashioned his small phallus to look more like a small clitoris. American intersex clinicians are extremely nervous about what this case might mean to their practices; a ruling in the child's favor could mean a new era of legal protections for these kinds of children and critical social acknowledgment of the harm wrought on generations of intersex children.

Meanwhile, in the United States, many teenage and adult transgender people still can't get the kinds of medical and surgical interventions that intersex babies get without ever asking. Fortunately, in the last ten years, there has been a movement toward providing better access to gender-affirming interventions for adolescents and adults who are transgender. More and more clinics at American children's hospitals are claiming to provide genuinely supportive care to this population.

In fact, the pendulum may be swinging too violently. In the name of being "affirming" of gender nonconforming children, some parents and clinicians now seem *too* quick to assume that such kids must be subject to gender reassignments that include nontrivial hormone regimens and surgical procedures. (Recall that most gender-nonconforming boys in clinical studies have grown up to be gay men, not transgender women. Many gender-nonconforming girls grow up to be satisfied with their birth gender assignments and bodies.) What looks at first like progress may sometimes amount to the same old rush to normalize "deviance" into a heterosexist two-sex vision of the world, rather than accepting that biology and identity are not the same thing. Calm, evidence-based care in the treatment of gender nonconforming children—care that doesn't rush to "resolve" their identities—is only now emerging. European clinicians again seem to be leading.

"Born with one sex's brain in the body of another sex, so needing rescue by the doctor"—that's still the transgender narrative that goes down smoothest among straight adults, perhaps especially among parents who long for a simple resolution to their seemingly challenging children. But in the United States, there is a growing movement of parents pushing to let gender-nonconforming children just be who they are, without diagnostic perusal and medical intervention. True radicals, these parents of "pink boys" and "blue girls" help other such parents learn to simply love without risky "normalizations," letting children grow up to decide for themselves what bodily changes, if any, they want to pursue.

I think that not too far in the future, this approach will no longer be seen as radical. American culture has made big strides since the intersex-rights movement began in 1993. Social acceptance of sex and gender variation has certainly increased dramatically in the last decade, and legal protections for transgender people, gay men, lesbian women, and bisexual people have been expanding. That said, many are still subject to housing, employment, and medical discrimination as well as to bullying and bashing. I'm working now on a committee assembled by the Association of American Medical Colleges to try to get medicine to lead in positive social change for people who are in the social minority in terms of their gender histories and sexual orientations. But as of now, American medicine remains a potential venue for oppression as much as for liberation.

There is much reasonable disagreement among transgender activists as to the right role for medicine in transgender politics. Clinicians who work with transgender people know that they are much more diverse in experiences, senses of self, and needs than the general public realizes. Clinicians with whom I speak sometimes express frustration that they have to toe particular party lines (like "transgender always means 'born trapped in the wrong body'") or risk being painted as anti-trans, even when they are struggling to put the needs and desires of a patient before politics.

As for Andrea James and Lynn Conway—the two trans women who most intensely went after Mike Bailey and then me—while they are well known among transgender activists, I think it is fair to say that they are not generally viewed as leaders of the rights movement. That kind of leadership comes more from groups like the National Center for Transgender Equality. James and Conway keep up their campaigns against Bailey, me, and pretty much anyone else aligned with Blanchard's understanding of male-to-female transsexualism as involving sexual orientation as well as gender identity.

My journal article on the Bailey book controversy still brings me frequent mail from men and transgender women who tell me the article helped them figure out that they are autogynephilic in sexual orientation. Some also write to tell me how wrong Blanchard and Bailey are, stating that their gender experiences have nothing to do with eroticism whatsoever or that they are a "third type" not captured by Blanchard's taxonomy. (Blanchard never said there couldn't be other types; he simply argued that sexual orientation was salient to requests for gender reassignment among natal males.) In spite of substantial pressures not to do so, in the last five years, more and more people have been openly self-identifying as autogynephilic; some identify mostly as male but live part-time as women, while others identify as transgender women. In 2013, Anne Lawrence, the physician-researcher who self-identifies as autogynephilic, published a groundbreaking book analyzing hundreds of autobiographical narratives of autogynephilia. The book encourages all readers to understand that, for those autogynephilic natal males who choose permanent social transition, the gender identity of female makes the most sense for their lives—that fully taking on a female identity is not a game, a fantasy, or a fetish, but a necessary means of survival, just as coming out is for gay and lesbian people. The work of Lawrence and others continues to show that transition greatly improves the psychological and social lives of transgender people who seek it and are given supportive care, including access to the desired hormonal and surgical interventions. Their lives are, on average, made safer and better.

Meanwhile, some researchers keep looking to see if they can find evidence that transgender people really do have the brain of one sex in the body of the other. Results of such work are mixed and difficult to analyze, in part because the sex-specific parts of the brain can change even in adulthood—for example, when someone starts taking cross-sex hormones. For my part, I hope we never require biological "proof" to believe someone's self-declaration of gender. As one wise guardian of a seven-year-old child with intersex once said to me, "We don't need another blood test to figure out who he is. He's already told us he's a boy." That, it seems to me, is all we ought to require of people as we "decide" what genders they are. How they got there may be scientifically interesting to us (it certainly is to me), but how they identify themselves as individual persons in terms of gender is for *them* to decide.

Anne Lawrence's career trajectory is typical of researchers who have been put through the wringer for making unpopular identity claims: Such people don't change course just because they've had the snot kicked out of them. Among all of the researchers whom I interviewed specifically because their work got them into trouble, not one has disavowed the controversial research. A lucky few have found their work now generally in favor. Elizabeth Loftus, for example, is now widely recognized as having been right about how fictional "memories" can be implanted in the human brain, even in some cases of alleged childhood sexual abuse. Ed Wilson's basic idea that there are evolutionary bases for human social behaviors is now more accepted among the general public and even academics.

Like Wilson, some researchers whom I met because of their intense political messes are now doing less controversial work, not because they're afraid of shitstorms but because they don't *need* them. (They lack the Galilean personality.) For example, since *A Natural History of Rape*, Craig Palmer has been studying social networks of Newfoundlanders and representations of rescuers in Holocaust museums. Chuck Roselli (the "gay sheep" guy) has continued to be interested in hormonal effects on the brain, but he's also been working on understanding how certain herbal preparations might help treat prostate cancer.

Among those *with* the Galilean personality, Mike Bailey has continued to study sexual orientation, collaborating on one project looking at the genetic basis of male homosexuality and on another aimed at understanding what sexual arousal looks like in the brains of bisexuals. Although age has taken the sharp edge off his style, life has not exactly quieted down for him. In early 2011, a couple of years after my work on his controversy came out, Mike managed to get himself in a whole new round of trouble by allowing a pair of exhibitionists to demonstrate something called a *fucksaw* in an after-class special for students in his Northwestern human sexuality class. In case you're not familiar with tools that require shopping at both Good Vibrations and Home Depot, a fucksaw is a do-it-yourself sex toy made by combining a reciprocating saw with a dildo. (At least that's how it has been described to me; I can't say I've had the pleasure.) News of this event leaked out into the local, then the national, then the international media. I heard about it accidentally via CNN.

When I got together with Mike over a tense drink to talk about it, I asked him what he was thinking to allow a live sex demonstration in the Ryan Family Auditorium on Northwestern's Evanston campus. He replied, "Well, Alice, the fucksaw couple asked me before *that* if they could do *fire play*"—sex games involving fire—"and I answered, 'No! That'll get me in trouble!'" Hearing this, I involuntarily smacked him in the head. The only upside was that Aron took to referring to Mike as the Fucksaw, simplifying certain family conversations, as we have three friends named Mike. Predictably, various academics tried to defend the fucksaw show by appealing to "academic freedom" and "freedom of speech." By Mike's own admission, however, he'd put no more than five minutes of thought into the decision to allow this demonstration. For an act to be protected by academic freedom, it really should have arisen from actual academic thought. Allowing the fucksaw demo was just dumb, and I think Mike ultimately agreed, though as usual he would say it was a dumb choice only because the general public is so incredibly stupid about sex and, as usual, he'd have a point. The students in Mike's class not only didn't object to the demonstration (which they'd been allowed to skip if they wished); many actively defended him and the course to the media. Nevertheless, the Northwestern University administration responded to the kerfuffle by taking the course away from Bailey. It is now taught by someone more politically correct.

(Not me.) Mike's son Drew has finished his PhD and is now an assistant professor studying the evolution of sex differences.

What about Napoleon Chagnon? In 2010, the journal *Human Nature* peer-reviewed and published my findings about what the American Anthropological Association did to aid and abet Patrick Tierney in smearing Chagnon and James Neel, and in the cycle immediately following publication of my exposé, Chagnon was finally elected to the National Academy of Sciences. Shortly after that, he finally finished his decade-in-the-making memoir, publishing it under the title *Noble Savages: My Life Among Two Dangerous Tribes—the Yanomamö and the Anthropologists.* Emerging from the northern Michigan woods to return to academia, in 2012 Chagnon accepted a new professorship at the University of Missouri–Columbia (in Craig Palmer's department). With the help of his colleague Ray Hames of the University of Nebraska, Chagnon is finally back at work processing the mountains of data he collected among the Yanomamö.

Several of the anthropologists I criticized in my work are still complaining that while I may have proved that Neel and Chagnon were wrongly accused of genocide, I haven't adequately investigated Chagnon's other putative ethical transgressions, such as allegedly manufacturing scenes for documentaries and cavorting with habitat-destroying gold miners. Were these charges coming from people who hadn't first cried genocide, I might be more inclined to pursue them. As it is, the AAA has yet to apologize to Chagnon for what happened under its auspices. Chagnon used a lawyer I recommended to him to finally force the AAA to heed the vote of its membership and take down from its Web site the El Dorado Task Force Report, which included the claim that Chagnon had paid his Yanomamö subjects to murder each other.

Today, the Yanomamö are in dire straights. As recently reported in *The Washington Post*, "a new bill pending before Brazil's Chamber of Deputies would proclaim a 'public interest' in allowing Indian reserves [like that on which the Yanomamö live] to be used for farming, mining, oil and gas pipelines, hydroelectric dams, human settlements and military operations." This seems guaranteed to end their way of life.

It's difficult for me to explain what has happened to the *American Journal of Bioethics* (*AJOB*), if only because I don't employ a research team to help me keep up with the journal's scandals. I guess I should relay at least the fact that Glenn McGee—the ethically challenged *AJOB* editor who remained unmoved by our calls for conflict-of-interest disclosures on the dex papers—ended up resigning from *AJOB* in early 2012 after some of his critics discovered that he had started working as an executive for a Texas stem-cell therapy company named Celltex. Celltex licensed the wares of a Korean stem-cell outfit called RNL Bio, a company McGee had been defending in the press a year earlier, after two of RNL's stem-cell patients had mysteriously died. Many folks in bioethics didn't think it all that funny that a stem-cell executive was now running a prominent medical ethics journal, especially because his company appeared to be running afoul of FDA regulations. Not long after McGee relinquished his editorship of *AJOB*, he also quit Celltex, which was soon under investigation by the FDA. Before he left *AJOB*, McGee saw his third wife, the bioethicist Summer Johnson McGee, installed as coeditor in chief of *AJOB* along with a friend of McGee's.

Summer herself resigned just a few months later, on the heels of a Senate inquiry into a pain-medication scandal in which *AJOB* was implicated. The journal had published a couple of major pro-pain-medication articles without including information about the authors' funding by the drugmakers whose wares they were advocating. (Yes, more undisclosed conflicts of interest.) Before Summer's departure, *AJOB* had essentially been forced to publish one of the longest post-publication conflict-of-interest disclosures anybody had ever seen. I never could convince the remaining editor—the McGees' friend David Magnus, a bioethicist at Stanford—to fix any of the problems with the dex papers. Robert "Skip" Nelson, the FDA official who ran the investigation on prenatal dex, still works as a paid ethicist for the FDA and as an *AJOB* editor in chief.

Dix Poppas, the Cornell medical school pediatric urological surgeon who was "sensory-testing" little girls' genitals after he surgically "feminized" them, is apparently still doing these surgeries. After we made complaints, Cornell successfully argued to the Office for Human Research Protections (OHRP) that Poppas's work should not be of concern to them because it was not federally funded and because he was supposedly doing the genital stimulation tests as part of "normal" medical care. (OHRP regulates federally funded research, not ordinary medical care.) In 2014, Poppas was voted one of *New York Magazine*'s Best Doctors.

As to prenatal dexamethasone, I still think it is highly likely that many of the pregnant women guided into the intervention by Maria New never really understood they were in a high-risk experiment. In the United States patients of physicians like Poppas and New—vulnerable people who are called "patients" while in the clinic and then used retrospectively as research subjects by the same physicians—seem to be caught in a never-never land in terms of rights and protections. Maria New is still on faculty at Mount Sinai School of Medicine and the Web site of the Maria New Children's Hormone Foundation continues to advertise prenatal dex as "safe for mother and child." New's long run of NIH funding seems to have finally come to a permanent end in 2008, about a year and a half before we raised our concerns to the Feds. I don't know why it ended. She has been pursuing a new form of prenatal testing that would allow earlier diagnosis of fetal sex, potentially making it possible to avoid exposing any males "accidentally" to prenatal dexamethasone. The doctor who blew the whistle on New's Cornell NIH research in 2003 now works at the University of Minnesota. She and her lawyers were awarded $877,000 from a $4.4 million settlement paid by Cornell.

Studies of prenatal glucocorticoids like dexamethasone continue to cause great concern. One recent laboratory study showed that fetal ovaries (obtained from aborted fetuses) that are treated with dexamethasone suffer cell death, a finding that suggests that female fetuses exposed to dexamethasone for CAH may grow up to be women with fertility problems. How painfully ironic given that one of the goals of New et al.'s intervention has been to make the females more likely to be mothers. A recent study from Finland shows an association between prenatal glucocorticoid exposure for risk of premature birth and negative mental-health effects, suggesting that "even at low dosages," glucocorticoids can have "a programming effect on the fetal brain." The authors of this study note that animal studies have also been pointing to this possibility.

Conspiracy theories are always tempting—they make one feel both clever and important—but there is little reason to believe that the feds did so little on prenatal dexamethasone only because of Robert "Skip" Nelson's ties to the *AJOB* gang. Thanks to the independent analysis of a dogged reporter named Theresa Defino, it's now clear that dex was no special case. The Office for Human Research Protections simply is no longer doing its job. While legitimate complaints have been arriving to OHRP's offices at a steady pace over the last several years, under the current directorship of Jerry Menikoff, OHRP has been opening fewer investigations, issuing fewer determination letters (findings of wrongdoing), and even lately backing off from enforcement requirements when researchers complain. According to Defino, "prior to 2009 [OHRP] typically had 20 open cases per year." But under Menikoff's leadership, in all of 2013, OHRP opened only one investigation. These days, OHRP seems to be much more about protection of research than protection of research subjects. (Several people who have recently quit OHRP have agreed with this reading when I have put it to them.) Regulations stand as they've been, but in practice, protections for people who become subjects of medical research may be at their weakest in decades.

I've now met quite a few researchers who, like me, stumbled on an ethical travesty they tried to stop, and like many of them, I wish I could let it go but feel that doing so would be irresponsible. I still have to remind myself regularly that the fact that we failed to get the federal government to act on dex doesn't mean we failed. The work my colleagues and I have done has made it impossible for a physician or a family doing basic Internet research on prenatal dexamethasone for CAH not to find out that it is controversial and involves troubling risks and unknowns. Of course, one would rather such information were made available more directly, so I'm still writing to textbook and journal editors, asking them to correct misrepresentations of the intervention and presenting my findings to physicians at medical schools. I also continue to track the history of dex by all the means available to me, including FOIA.

When I found out that prenatal dex was being used on a hunch with the goal of miscarriage prevention by IVF doctors, I relayed the news to the women at DES Action, an activist group I had come to know via my dex work. Recall that DES had also been used with the aim of preventing miscarriage. The women at DES Action were as stunned as I. A leader in the group, Kari Christianson (herself a woman prenatally exposed to DES), responded: "Again and again, unproven and unsafe drugs are available, offered and given to pregnant women without fully informed consent or understanding at a most vulnerable time. It is unconscionable. To think that we have learned little or nothing in the over forty years since DES health harm was brought to light is frightening beyond all reason."

The problem here is an old one: People don't really get that good intentions can't save you from hell. So long as we believe that bad acts are committed only by evil people and that good people do only good, we will fail to see, believe, or prevent these kinds of travesties. Nowadays I feel as though 90 percent of my time talking to academics and activists is spent trying to convince them of this: The people who are against you are not necessarily evil, and your own acts are not necessarily good. That's why we still need both scholars and activists. It's not easy to see what's what in the heat of the moment, and we need people pushing for truth *and* for justice if we're going to get both right.

But most people I run into aren't like us historians. Most people I meet seem convinced that the goodness of their souls will keep them from committing bad acts. When they look back at history, they don't see what we historians see—dumb tragedies. They see simple moral dramas, with predictable characters enacting easy stories of good and evil. They don't understand that the Nazis probably didn't think they were "Nazis."

Everybody knows the most famous line about history—Santayana's "Those who cannot remember the past are condemned to repeat it." But if this project has taught me anything, it's this: People don't want to listen to us historians and our warnings. People don't believe us when we come in at the start or the middle. They believe us only way after the end, if then. I'm learning to accept the fact that we are almost always too late. We can bear witness afterward, of course. And witnessing matters. But so many days, I find myself selfishly wishing that witnessing felt like enough.

ACKNOWLEDGMENTS

This book, which revolves around a meditation on a scientist whose work depended in large part on patrons, would have itself been impossible had I not had a faithful patron who has financially supported my work since I gave up tenure in 2004. Aron Sousa has supplied not only our son's college fund, two roofs (our house and my writing cottage), three squares, and cash for my lawyers and research trips, he has also provided life-saving companionship, laundry and cooking, editing of unclear drafts and crazy thoughts, a wonderful second family, and a consistent mandate that I should do what matters instead of what pays. I only wish my friends would stop asking what I've done to deserve him. (I have no answer.)

In 2008, the John Simon Guggenheim Foundation awarded me a fellowship to work on this book. I think that if they had not, I would never have had the guts to attempt this particular project. Moreover, if they had expected me, as many funders do, to complete the project in one year, I would never have learned what I have. I will always be grateful to them for the vote of confidence and the long view of productivity, and grateful to the people who supported my Guggenheim application: Barron Lerner, Steve Pinker, Dan Savage, and Miriam Schuchman.

For three years, this project was supported by a human sexualities grant from the Provost's Office of Northwestern University. I appreciate that financial support, as well as the administrative support given to me at Northwestern by Dan Linzer, Ray Curry, Kathryn Montgomery, and Tod Chambers. Ray and Kathryn in particular offered extraordinary amounts of psychological support, the kind you have to have to survive doing the kind of work I do in academia today. My colleagues and the staff in the Medical Humanities and Bioethics Program at Northwestern have been nothing short of amazing; thanks to Kathryn and Tod, Catherine Belling, Gretchen Case, Megan Crowley-Matoka, Sydney Halpern, Kristi Kirschner, Myria Knox, Bryan Morrison, Debjani Mukherjee, Sarah Rodriguez, and Katie Watson. The staff of the Galter Library has been tremendously helpful in tracking down difficult to locate texts.

Over the years of my work, starting in graduate school, additional funding has been generously provided to me by the Woodrow Wilson Foundation, Indiana University, Michigan State University, Cornell College, the California Endowment, the Gill Foundation, the Arcus Foundation, the National Endowment for the Humanities (through the Hastings Center), the Social Sciences and Humanities Research Council of Canada (through the Enhancement Technologies and Human Identity Working Group), and the Canadian Institutes of Health Research (through Impact Ethics: Making a Difference).

The Impact Ethics group has been critically important to my thinking processes in the last two years. I am especially grateful to Françoise Baylis, Jocelyn Downie, Barry Hoffmaster, and Leigh Turner. For help in thinking through and sometimes cowriting about the Bailey, Tierney, and/or dexamethasone projects, as well as pediatric normalizations, academic freedom, and research ethics, I am particularly grateful to Marc Breedlove, Jim Brown, Andrew Burnett, Ann Carmichael, Mike Carome, MK Czerwiec, Carl Elliott, Joel Frader, Ed Goldman, Phil Gruppuso, the late Mary Ann Harrell, Kelly Hills, Mark Hochhauser, Joel Howell, Cindy Jordan, Christine Kelsey, Rosa Lee Klaneski, Anne Lawrence, Hilde Lindemann, Ruth Macklin, Jamie Nelson, Nigel Paneth, Erik Parens, Bill Peace, Susan Reverby, John Schwartz, Lois Shepherd, Jason Stallman, the late Kiira Triea, Eric Vilain, Roger Webb, and Sid Wolfe. Research assistance for projects discussed in this book came from a delightful series of people who were or then became friends: Colleen Kiernan, Taylor Sale, Yorgos Strangas, and Val Thonger. For legal advice and representation, I am grateful to Karen Mayer at Penguin Press, Cathy Jacobs, Thad Morgan, and J. J. Burcham. Janet Green and Anne Tamar-Mattis turned out to be ideal company in the dexamethasone forest, and for her classy public leadership through that forest, I am also grateful to Hilde Lindemann. Ken Kipnis also deserves special mention, for making me both laugh and cry every time he wished me "strength to your arm." My thanks also go to all the people who provided interviews and other source material, including especially Mary McCarthy, and all of the people who reviewed my manuscripts on these topics, including especially the journal editors Randy Cruz, Jane Lancaster, and Ken Zucker. I am also indebted to Doug Hume for his online "AnthroNiche" document collection.

I hope the text of this book at least hints at the gratitude I feel toward my parents for bringing me up with a sense of purpose. What I am sure is not clear in this book is how my siblings (including my "sisish") and my friends have kept me sane and even laughing. In terms of "chosen" family, I am particularly grateful to Val Thonger and Ken Sperber, Libby Bogdan-Lovis and Bill Lovis, Danny Black, April G. K., April Herndon, Vic Loomis, Ellen Weissbrod, and Paul Vasey.

Perhaps the best thing about this work is the three close friends the three major projects in this book have brought me: Ray Blanchard, Ray Hames, and Ellen Feder. If Ellen had not been by my side through the entire dexamethasone affair, I think I might not have been able to continue that work. Her clarity, humility, and anger formed a guiding light for me. David Sandberg was already a friend when he pulled me into dex but, through that work, he became someone with whom I look forward to growing old and more crotchety.

Mark Oppenheimer did me the great good service of introducing me to Betsy Lerner, who was kind enough to become my agent and to be the person who shaped this project into something that a mainstream press would understand. She then arranged a contract with editor Colin Dickerman at Penguin Press, a dream come true. After Colin left, Jeff Alexander became the chief editor for the book, and, with the help of editorial assistants Sofia Groopman and Will Carnes, brought to the work the focus and clarity it had been needing. I cannot imagine a better agent-editor combination than Betsy and Jeff.

Finally, I would like to thank our son for his interest in and support of my work. It cannot have been easy to have grown up with this book, but somehow he has always managed to second his father's vote of confidence in the work. I am so thankful to him for the way he has raised me.

NOTES

INTRODUCTION: THE TALISMAN

4 **Ruth Hubbard:** See Ruth Hubbard, *The Politics of Women's Biology* (Rutgers, NJ: Rutgers University Press, 1990).

4 **Londa Schiebinger and Cynthia Eagle Russett:** Londa Schiebinger, *The Mind Has No Sex? Women in the Origins of Modern Science* (Cambridge, MA: Harvard University Press, 1989); Cynthia Eagle Russett, *Sexual Science: The Victorian Construction of Womanhood* (Cambridge, MA: Harvard University Press, 1989).

4 **Stephen Jay Gould:** Stephen Jay Gould, *The Mismeasure of Man* (New York: W.W. Norton, 1981). Gould's treatment of the work of Samuel George Morton's craniometry in that book has since come under significant criticism; see Jason E. Lewis et al., "The Mismeasure of Science: Stephen Jay Gould Versus Samuel George Morton on Skulls and Bias," *PLoS Biology* 9 (2011): e1001071.

6 **This article mapped out:** Alice Domurat Dreger, "Doubtful Sex: The Fate of the Hermaphrodite in Victorian Medicine," *Victorian Studies* 38, no. 3 (Spring 1995): 335–70. I later published my first book based on this work: Alice Domurat Dreger, *Hermaphrodites and the Medical Invention of Sex* (Cambridge, MA: Harvard University Press, 1998).

6 **Bo got my work:** On our early meeting and collaboration, see Alice Dreger, "Cultural History and Social Activism: Scholarship, Identities, and the Intersex Rights Movement," in *Locating Medical History: The Stories and Their Meanings*, ed. Frank Huisman and John Harley Warner (Baltimore: Johns Hopkins University Press, 2004): 390–409.

7 **someday experience orgasm:** I later documented and criticized this system in Alice Domurat Dreger, "'Ambiguous Sex'—or Ambivalent Medicine? Ethical Problems in the Treatment of Intersexuality," *Hastings Center Report* 28, no. 3 (1998): 24–35.

7 **clitoris had been amputated:** See Cheryl Chase, "Affronting Reason," in *Looking Queer: Body Image and Identity in Lesbian, Bisexual, Gay, and Transgender Communities*, ed. Dawn Atkins (Binghamton, NY: Harrington Park Press, 1998): 205–20; Elizabeth Weil, "What If It's (Sort of) a Boy and (Sort of) a Girl?" *New York Times*, Sept. 24, 2006, http://www.nytimes.com/2006/09/24/magazine/24intersexkids.html.

9 **Bailey had suggested:** J. Michael Bailey, *The Man Who Would Be Queen: The Science of Gender-Bending and Transsexualism* (Washington, DC: Joseph Henry Press, 2003).

9 **account of the controversy:** My article appeared as Alice Dreger, "The Controversy Surrounding *The Man Who Would Be Queen:* A Case History of the Politics of Science, Identity, and Sex in the Internet Age," *Archives of Sexual*

Behavior 37, no. 3 (June 2008): 366–421, http://link.springer.com/article/10.1007%2Fs10508-007-9301-1.

10 ***New York Times*:** This was covered in Benedict Carey, "Criticism of a Gender Theory, and a Scientist Under Siege," *New York Times*, Aug. 21, 2007, www.nytimes.com/2007/08/21/health/psychology/21gender.html.

12 **Galileo actively argued:** This account of Galileo is based largely on the excellent biography by David Wootton, *Galileo: Watcher of the Skies* (New Haven, CT: Yale University Press, 2010).

13 **"cause of the eggs hardening":** Ibid., 164.

15 **"making human beings seem insignificant":** Ibid., 169.

16 **Founding Fathers were science geeks:** See Jonathan Lyons, *The Society for Useful Knowledge: How Benjamin Franklin and Friends Brought the Enlightenment to America* (New York: Bloomsbury Press, 2013).

17 **Galileo's middle finger:** See Rachel Donadio, "A Museum Display of Galileo Has a Saintly Feel," *The New York Times*, July 22, 2010, http://www.nytimes.com/2010/07/23/world/europe/23galileo.html.

17 **"Tiphaeus ever reached":** Translation from anonymous, "The Right Kinds of Relics," http://friendsofdarwin.com/2007/04/20070415/.

CHAPTER ONE: FUNNY LOOKING

19 **genital appearance upsets or worries some adult:** The somewhat shocking and non-evidence-based clinical pediatric approaches as they existed in the 1990s were documented in an important critical analysis by two gynecologists: Sarah Creighton and Catherine Minto, "Managing Intersex," *BMJ [British Medical Journal]* 323, no. 7324 (2001): 1264–65.

20 **Winston Churchill:** Speaking of Chamberlain, Churchill said, "Poor Neville, he will come badly out of history. . . . I know, because I will write the history." Quoted on p. 11 of Robert J. Caputi, *Neville Chamberlain and Appeasement* (London, England: Associated University Presses, 2000).

21 **about one in two thousand babies:** The medical literature contains no good study of the frequency of "ambiguous genitalia" (again, presumably because one would have to simply decide what would count). In order to get at an estimate, Bo Laurent (Cheryl Chase) and I asked specialists to tell us how often their teams were called to a birth because a baby's genitals were too unclear to assign a sex, and the figure consistently came to about one in fifteen hundred to one in two thousand.

21 **About one in three hundred babies:** This would include, for example, when a girl is born with a larger than expected clitoris or when a boy is born with hypospadias, i.e., when the urinary opening is not at the tip of the penis. The frequency of hypospadias is given in one current textbook as ranging from "between 0.4 to 8.2 cases per 1000 newborn boys"; see Bernardita Troncoso and Pedro-Jose Lopez, "Hypospadias," in *Pediatric Urology Book*, ed. Duncan Wilcox, Prasad Godbole, and Christopher Cooper, http://www.pediatricurologybook.com/hypospadias.html.

21 **one in a hundred:** Melanie Blackless et al., "How Sexually Dimorphic Are We? Review and Synthesis," *American Journal of Human Biology* 12, no. 2 (2000): 151–66.

21 **twenty-five and in graduate school:** In 2003, at the request of two historians of medicine editing a book on our profession, I wrote about why I became an activist-historian; see Alice Dreger, "Cultural History and Social Activism: Scholarship, Identities, and the Intersex Rights Movement," in *Locating Medical History: The Stories and Their Meanings*, ed. Frank Huisman and John Harley Warner (Baltimore: Johns Hopkins University Press, 2004): 390–409.

22 **one of my dissertation directors:** Because it bridged the history of medicine and science, my dissertation was codirected by Fred Churchill (historian of science) and Ann Carmichael (historian of medicine); see Alice Domurat Dreger, *Doubtful Sex, Doubtful Status: Cases and Concepts of Hermaphroditism in France and Britain, 1868–1915* (PhD dissertation, Indiana University, 1995).

23 **my three hundred primary sources:** I discuss this methodology and subject more fully in the book based on my dissertation: Alice Domurat Dreger, *Hermaphrodites and the Medical Invention of Sex* (Cambridge, MA: Harvard University Press, 1998).

23 **nineteenth-century Frenchwoman:** See the story of Louise-Julia-Anna in Ibid., 110–13, 138.

24 **feminine breasts with a penis:** We now know that mixed external sex anatomy can arise from a large number of conditions, including congenital adrenal hyperplasia in genetic females, partial androgen insensitivity syndrome in genetic males, various tumors, and polycystic ovary syndrome, just to name a few. Not all of the causes of mixed external sex anatomy are congenital (inborn).

24 **the other sex's organs inside:** We now understand that complete androgen insensitivity syndrome (cAIS) can cause a person to develop externally and behaviorally like a typical female, although internally she will have testes and will lack female reproductive organs (except for the vagina and vulva). It is not uncommon for this condition to go undiagnosed until late adolescence. We also now know that extreme forms of congenital adrenal hyperplasia (CAH) can cause a genetic female to develop as a fairly typical male in terms of external genitalia, so that the child would ordinarily be assumed to be male at birth, even though internally the child will have ovaries and a uterus.

24 **manly at puberty:** These would represent cases of 5-alpha-reductase deficiency, which causes a child to be born looking much like a typical female but to undergo a male-typical puberty. The protagonist of the novel *Middlesex* has this condition, as probably did Herculine Barbin. See Jeffrey Eugenides, *Middlesex: A Novel* (New York: Farrar, Straus & Giroux, 2002) and Michel Foucault, *Herculine Barbin: Being the Recently Discovered Memoirs of a Nineteenth-Century French Hermaphrodite*, trans. Richard McDougall (New York: Pantheon: 1980).

25 **the doctors' eyebrows rise:** These cases are traced in Dreger, *Hermaphrodites*.

25 **Age of Gonads:** This history is spelled out more fully in chapter 5 of Dreger, *Hermaphrodites*, and in Alice Dreger, "Hermaphrodites in Love: The Truth of the Gonads," in *Science and Homosexualities*, ed. Vernon A. Rosario (New York: Routledge, 1997): 46–66.

26 **Together Wilkins and Money:** This is best described in Sandra Eder, *The Birth of Gender: Clinical Encounters with Hermaphroditic Children at Johns Hopkins, 1940–1956* (PhD dissertation, Johns Hopkins University, 2011). See also Sandra Eder, "From 'Following the Push of Nature' to 'Restoring One's Proper Sex':

Cortisone and Sex at Johns Hopkins's Pediatric Endocrinology Clinic," *Endeavour* 36, no. 2 (2012): 69–76.

26 **sometimes lies:** For a collection of first-person accounts of this treatment system from people born intersex, see Alice Domurat Dreger, ed., *Intersex in the Age of Ethics* (Hagerstown, MD: University Publishing Group, 1999).

26 **core group:** For raw footage of intersex adults talking in the mid-1990s about what happened to them, see the videotape made by Bo Laurent/Cheryl Chase, *Hermaphrodites Speak!* (San Francisco: Intersex Society of North America, 1997), 30 minutes, http://www.youtube.com/watch?v=BwSOngdR7kM.

27 **Intersex Society of North America:** Bo Laurent (ISNA's founder) has written extensively on the motivations and origins of the intersex rights movement, often under her activist name, Cheryl Chase. See, for example, Cheryl Chase, "Affronting Reason," in *Looking Queer: Image and Identity in Lesbian, Bisexual, Gay and Transgender Communities*, ed. Dawn Atkins (Binghamton, NY: Harrington Park Press, 1998): 205–20; and Cheryl Chase, "Hermaphrodites with Attitude: Mapping the Emergence of Intersex Political Activism," *GLQ: A Journal of Lesbian and Gay Studies* 4, no. 2 (1998): 189–211.

27 **a few people did:** In the late nineteenth century, thanks to advances in anesthesia and infection control, surgery became safer and less painful, and at this point, a small number of hermaphroditic patients inquired about surgical options. I track this in Dreger, *Hermaphrodites*.

27 **while most seemed fairly unconcerned:** This variation is traced in Dreger, *Hermaphrodites*, but was first hinted at in the article that caused Bo to contact me: Alice Domurat Dreger, "Doubtful Sex: The Fate of the Hermaphrodite in Victorian Medicine," *Victorian Studies* 38, no. 3 (1995): 335–70.

27 **Bo was to be counted:** Bo's personal history was recounted in various documentaries as well as in Elizabeth Weil, *"What If It's (Sort of) a Boy and (Sort of) a Girl?" New York Times*, Sept. 24, 2006, www.nytimes.com/2006/09/24/magazine/24intersexkids.html.

28 **Bo and I later successfully worked to get rid of it:** The article where we argued for the change in nomenclature is: Alice D. Dreger, Cheryl Chase, Aron Sousa, Philip A. Gruppuso, and Joel Frader, "Changing the Nomenclature/Taxonomy for Intersex: A Scientific and Clinical Rationale," *Journal of Pediatric Endocrinology and Metabolism* 18, no. 8 (2005): 729–33. Shortly thereafter, the medical establishment officially dropped all diagnoses based on the term "hermaphrodite" and adopted the umbrella term "disorders of sex development" for all intersex conditions; see Peter A. Lee et al., "Consensus Statement on Management of Intersex Disorders" (also known as the Chicago Consensus), *Pediatrics* 118 (2006): e488–e500. This shift was controversial among activists; see Ellen K. Feder, "Imperatives of Normality: From 'Intersex' to 'Disorders of Sex Development,'" *GLQ: A Journal of Lesbian and Gay Studies* 15, no. 2 (2009): 225–47; Georgiann Davis, *The Dubious Diagnosis: How Intersex Became a Disorder of Sex Development* (New York: New York University Press, 2015).

28 **Bo had also been born with ambiguous genitalia:** See Weil, "What If . . ."

28 **marshaled her lesbian feminist political consciousness:** See Chase, "Affronting Reason," and Chase, "Hermaphrodites with Attitude."

29 **took on a new name:** See "Cheryl Chase (Bo Laurent)," www.isna.org/about/chase.

30 **latest medical books:** For documentation and analyses of the homophobia behind the modern medical management of intersex, see Alice Domurat Dreger, "'Ambiguous Sex'—or Ambivalent Medicine? Ethical Problems in the Treatment of Intersexuality," *Hastings Center Report* 28, no. 3 (1998): 24–35; Suzanne J. Kessler, *Lessons from the Intersexed* (New Brunswick, NJ: Rutgers University Press, 1998); Anne Fausto-Sterling, *Sexing the Body: Gender Politics and the Construction of Sexuality* (New York: Basic Books, 2000).

30 **name in the snow:** Adrienne Carmack, Lauren Notini, and Brian D. Earp, "Should Elective Surgery for Hypospadias Be Performed before an Age of Consent?," forthcoming, includes this typical medical construction of the problem: "It is the inalienable right of every boy to be a pointer instead of a sitter by the time he starts school and to write his name legibly in the snow"; from O. S. Culp and J. W. McRoberts, "Hypospadias," in C. E. Alken, V. Dix, and W. E. Goodwin, eds., *Encyclopedia of Urology* (New York: Springer-Verlag, 1969): 11307–44.

31 **Martha Coventry:** See Martha Coventry, "Finding the Words," in Dreger, *Intersex in the Age of Ethics*, 71–76.

31 **David Cameron Strachan:** See David Cameron, "Caught Between: An Essay on Intersexuality," in Dreger, *Intersex in the Age of Ethics*, 91–96; David Cameron, "Being Different and Fitting In," http://oiiinternational.com/2538/fitting; Anonymous, "2008 LGBT Heroes: David Cameron Strachan, Intersex Community Volunteer Activist," http://www.kqed.org/community/heritage/lgbt/heroes/2008.jsp.

32 **gazing upon her in the book:** For more on the medical display of intersex people, see Sarah Creighton et al., "Medical Photography: Ethics, Consent and the Intersex Patient," *BJU International* 89, no. 1 (2002): 67–71; and see Alice Domurat Dreger, "Jarring Bodies: Thoughts on the Display of Unusual Anatomies," *Perspectives in Biology and Medicine* 43, no. 2 (2000): 161–72.

33 **daughter's noticeably long clitoris:** Bo and I interviewed this mother and daughter on the record in "A Mother's Care: An Interview with 'Sue' and 'Margaret,'" in Dreger, *Intersex in the Age of Ethics*, 83–89.

33 **her clitoris was bigger than most:** This woman provided a short essay for an anthology I collected; see Kim, "As Is," in Dreger, *Intersex in the Age of Ethics*, 99–100.

33 **she had testes inside:** We alluded to this story in Dreger et al., "Changing the Nomenclature."

33 **a uterus inside of him:** This man was a genetic female with an extreme form of congenital adrenal hyperplasia.

34 **now she was going to die:** Bo and I often showed people a surgical training video that explained that sometimes "for social reasons" surgeons "needed" to shorten clitorises on very young babies, before it was really medically advisable to attempt anesthesia; see Richard Hurwitz et al., "Surgical Reconstruction of Ambiguous Genitalia in Female Children," (Woodbury, CT: Cine-Med, 1990).

34 **Bruce Wilson:** See Bruce E. Wilson and William G. Reiner, "Management of Intersex: A Shifting Paradigm," in Dreger, *Intersex in the Age of Ethics*, 119–35.

34 **"phall-o-meters":** The phall-o-meters were inspired by an article by Suzanne Kessler, "Meanings of Genital Variability," *Chrysalis: The Journal of Transgressive Gender Identities* 2 (1997): 33–38.

34 **fit social norms:** For more on medical interpretations of "correct" phallus size, see Dreger, *Hermaphrodites*, 183.

35 **extensive ethical critique:** See Dreger, "'Ambiguous Sex.'"

35 **next book I published:** See Dreger, *Intersex in the Age of Ethics*; this was based on a special journal issue on intersex, *Journal of Clinical Ethics* 9, no. 4 (Winter 1998).

35 **I paid a university photographer:** I explained the logic behind this act in Dreger, "Jarring Bodies."

36 **As *Nature Made Him*:** John Colapinto, *As Nature Made Him: The Boy Who Was Raised as a Girl* (New York: HarperCollins, 2000).

37 **Reimer also failed to prove Money's theory:** This story first broke in 1997, startling the medical establishment, but did not garner widespread public attention until Colapinto's treatment. See Natalie Angier, "Sexual Identity Not Pliable After All, Report Says," *New York Times*, Mar. 14, 1997, http://www.nytimes.com/1997/03/14/us/sexual-identity-not-pliable-after-all-report-says.html; this was a front-page story on an academic journal report from critics of Money; Milton Diamond and H. Keith Sigmundson, "Sex Reassignment at Birth: Long-Term Review and Clinical Implications," *Archives of Pediatric and Adolescent Medicine* 151, no. 3 (1997): 298–304.

38 **most intersex people kept the gender assignments:** Gender outcomes are reviewed in Lee et al., "Consensus Statement."

38 **Bo said it as plainly as she could:** See Cheryl Chase, "What Is the Agenda of the Intersex Patient Advocacy Movement?" *Endocrinologist* 13, no. 3 (2003): 240–42. See also Lee et al., "Consensus Statement," for evidence that the clinical establishment was by 2006 recognizing Bo's and ISNA's formulation of the problem: "Although clinical practice may focus on gender and genital appearance as key outcomes, stigma and experiences associated with having a DSD [disorder of sexual development] (both within and outside the medical environment) are more salient issues for many affected people" (p. e496).

38 **Bill Reiner:** For Reiner's challenges to John Money's established paradigm, see Wilson and Reiner, "Management of Intersex," and see William G. Reiner and John P. Gearhart, "Discordant Sexual Identity in Some Genetic Males with Cloacal Exstrophy Assigned to Female Sex at Birth," *New England Journal of Medicine* 350, no. 4 (2004): 333–41.

39 **maximin strategy:** Howard Brody and James R. Thompson, "The Maximin Strategy in Modern Obstetrics," *Journal of Family Practice* 12, no. 6 (1981): 977–86.

39 **resulted in *more* net harm:** For a synopsis of this ongoing problem in obstetrics, see Aron C. Sousa and Alice Dreger, "The Difference between Science and Technology in Birth," *JAMA Virtual Mentor* 15, no. 9 (2013): 786–90.

39 **the founder of pediatric endocrinology:** See Eder, *The Birth of Gender.*

40 **Money had *known*:** See Colapinto, *As Nature Made Him.*

40 **Articles and op-eds:** See, for example, Creighton and Minto, "Managing Intersex"; Wilson and Reiner, "Management of Intersex" (originally published in *Journal of Clinical Ethics* 9, no. 4 [1998]: 360–69); Kenneth Kipnis and Milton

Diamond, "Pediatric Ethics and the Surgical Assignment of Sex," *Journal of Clinical Ethics* 9, no. 4 (1998): 398–410 (republished in Dreger, *Intersex in the Age of Ethics*, 173–193).

41 **conjoined twins:** This is explained in Alice Dreger, *One of Us: Conjoined Twins and the Future of Normal* (Cambridge, MA: Harvard University Press, 2004). See also Alice Dreger, "The Sex Lives of Conjoined Twins," *The Atlantic* (Oct. 25, 2012), www.theatlantic.com/health/archive/2012/10/the-sex-lives-of-conjoined-twins/264095.

41 **UCLA surgeon:** See Dreger, *One of Us*, 62.

41 **more impairment and shorter life spans:** See Dreger, *One of Us*. See also Alice Domurat Dreger, "The Limits of Individuality: Ritual and Sacrifice in the Lives and Medical Treatment of Conjoined Twins," *Studies in History and Philosophy of Biological and Biomedical Sciences* 29c, no. 1 (Mar. 1998): 1–29. See also Alice D. Dreger and Geoffrey Miller, "Conjoined Twins" in *Pediatric Bioethics*, ed. Geoffrey Miller (Cambridge, UK: Cambridge University Press, 2010): 203–18.

42 **Ladan and Laleh Bijani:** see Dreger, *One of Us*, 41–43 and 66–67.

42 **"achieve their dream of separation":** Keith Goh quoted in Anonymous, "Nation in Shock over Death of Iranian Twins," *Belfast News Letter* (Northern Ireland), July 9, 2003, 14.

42 **political consciousness about LGBT:** See Chase, "Hermaphrodites with Attitude."

43 **discrimination against a sexual minority:** On this point, see Alice D. Dreger and April M. Herndon, "Progress and Politics in the Intersex Rights Movement: Feminist Theory in Action," *GLQ: A Journal of Lesbian and Gay Studies* 15, no. 2 (2009): 199–224.

43 **penetrated by men:** For a review of the evidence of homophobia in the medical literature on intersex, see Dreger, "'Ambiguous Sex'"; Kessler, *Lessons from the Intersexed*; Fausto-Sterling, *Sexing the Body*.

43 **"I'm not a doctor":** I wrote about this technique of relationship-building in Alice Dreger, "Sleeping with the Enmity," *Atrium*, no. 3 (2006): 12.

46 **Oprah:** See "Growing Up Intersex," *The Oprah Winfrey Show*, July 19, 2008, www.oprah.com/oprahshow/Growing-Up-Intersex.

47 **Richard Rink:** The press release I wrote for ISNA on this was published as Alice Dreger, "Urologists: Agonize over Whether to Cut, Then Cut the Way I'm Telling You," Intersex Society of North America, Oct. 14, 2004, www.isna.org/articles/aap_urology_2004.

48 **San Francisco Human Rights Commission:** For the report that emerged, see Marcus de María Arana, ed., *A Human Rights Investigation into the Medical "Normalization" of Intersex People*, a Report of a Public Hearing by the Human Rights Commission (City and County of San Francisco, 2005).

48 **wrote something like this:** See Weil, "What If . . ."

49 **simply to be treated as human:** See Alice Dreger, "Intersex and Human Rights: The Long View," *Ethics and Intersex*, ed. Sharon E. Sytsma (Doetinchem, Netherlands: Springer, 2006), 73–86.

CHAPTER 2: RABBIT HOLES

50 **two handbooks:** These were compiled in their first form by Sallie Foley and Christine Feick, and ultimately published in 2006 as *Clinical Guidelines for the Management of Disorders of Sex Development in Childhood* and *Handbook for Parents* (now available through Accord Alliance, www.accordalliance.org/dsd-guidelines/).

51 **list of talking points:** I wrote about this strategy in Alice Dreger, "Footnote to a Footnote: On Roving Medicine," in *Bioethics Forum*, Oct. 9, 2008, www.thehastingscenter.org/Bioethicsforum/Post.aspx?id=2484.

51 **international medical consensus:** See Peter A. Lee et al., "Consensus Statement on Management of Intersex Disorders" (also known as the Chicago Consensus), *Pediatrics* 118 (2006): e488–e500

51 **real problem in intersex care:** See for example, Richard S. Hurwitz, "Long-Term Outcomes in Male Patients with Sex Development Disorders—How Are We Doing and How Can We Improve?," *Journal of Urology* 184, no. 3 (2010): 821–32.

53 **study of "fag hags":** Nancy H. Bartlett, H. M. Patterson, Douglas P. VanderLaan, and Paul L. Vasey, "The Relation Between Women's Body Esteem and Friendships with Gay Men," *Body Image* 6, no. 3 (2009): 235–41.

53 **Bailey transsexualism controversy:** Alice Dreger, "The Controversy Surrounding *The Man Who Would Be Queen:* A Case History of the Politics of Science, Identity, and Sex in the Internet Age," *Archives of Sexual Behavior* 27, no. 3 (2008): 366–421, http://link.springer.com/article/10.1007%2Fs10508-007-9301-1.

54 **popular, comforting narrative:** For an example of this kind of narrative of transgender, see Randi Ettner, *Confessions of a Gender Defender: A Psychologist's Reflections on Life Among the Transgendered* (Chicago: Spectrum Press, 1996).

54 **quest for the true self:** Carl Elliott discussed the connection between the standard story of transsexualism and American narratives of authenticity in *Better Than Well: American Medicine Meets the American Dream* (New York: Norton, 2003).

55 **This rankled Bailey:** Bailey told me that it was Ettner's book, *Confessions of a Gender Defender,* that made him determined to write the "true" story in a book of his own; J. Michael Bailey interview with Alice Dreger, Aug. 8, 2006, revised transcript received Aug. 8, 2006, and e-mail from Bailey to Dreger, Aug. 22, 2006. This is also discussed in Dreger, "Controversy," 371.

55 ***The Man Who Would Be Queen:*** J. Michael Bailey, *The Man Who Would Be Queen: The Science of Gender-Bending and Transsexualism* (Washington, DC: Joseph Henry Press, 2003).

55 **advice of colleagues:** Ray Blanchard, interview with Alice Dreger, Aug. 2, 2006, revised transcript received Aug. 3, 2006; also discussed in Dreger, "Controversy," 377.

56 **"becoming a girl":** Bailey, *Man Who Would Be Queen*, 50.

56 **Blanchard concluded:** See Ray Blanchard, "The Concept of Autogynephilia and the Typology of Male Gender Dysphoria," *Journal of Nervous and Mental Disease* 177 (1989): 616–23.

58 **"well suited to prostitution":** Bailey, *Man Who Would Be Queen*, 185.

59 **Blanchard coined a new term:** See Blanchard, "Concept of Autogynephilia."

60 ***fa'afafine*:** Paul L. Vasey and Nancy H. Bartlett, "What Can the Samoan 'Fa'afafine' Teach Us About the Western Concept of Gender Identity Disorder in Childhood?" *Perspectives in Biology and Medicine* 50, no. 4 (Autumn 2007): 481–90.

61 **Richard/Alice Novic:** See Richard J. Novic, *Alice in Genderland: A Crossdresser Comes of Age* (iUniverse, 2009).

62 **well-screened trans women:** See, for example: Ray Blanchard, "Gender Dysphoria and Gender Reorientation," in B. W. Steiner, ed., *Gender Dysphoria: Development, Research, Management* (New York: Plenum Press, 1985): 365–92; Ray Blanchard, "The Case *for* Publicly Funded Transsexual Surgery," *Psychiatry Rounds* 4, no. 2 (Apr. 2000): 4–6. See also Dreger, "Controversy," 415.

62 **sex reassignment in Canada:** Alice D. Dreger, "Response to the Commentaries on Dreger (2008)," *Archives of Sexual Behavior* 37 (2008): 503–10; see 504.

63 **"become the women they love":** Bailey, *Man Who Would Be Queen*, p. xii.

63 ***paraphilic*:** Ibid., 171–72.

63 **gatekeepers for sex reassignment:** See Joanne Meyerowitz, *How Sex Changed: A History of Transsexuality in the United States* (Cambridge, MA: Harvard University Press, 2004).

64 **In 1969, one clinician:** See Howard J. Baker, "Transsexualism: Problems in Treatment," *American Journal of Psychiatry* 125 (1969): 1412–18.

64 **Paul McHugh:** See Paul McHugh, "Transgender Surgery Isn't the Solution," *Wall Street Journal*, June 12, 2014, http://online.wsj.com/articles/paul-mchugh-transgender-surgery-isnt-the-solution-1402615120.

64 **Bailey actually criticizes in his book:** Bailey, *Man Who Would Be Queen*, 207.

64 **liposuction on anorexics:** See Paul R. McHugh, "Psychiatric Misadventures," www.lhup.edu/~dsimanek/mchugh.htm (accessed July 26, 2014).

64 **"appropriating this body for themselves":** Janice G. Raymond, *The Transsexual Empire: The Making of the She-Male* (Boston: Beacon Press, 1979), 104. A new book by feminist Sheila Jeffreys has revised this debate over the relationship of feminism to transgender; see Michelle Goldberg, "What Is a Woman?" *The New Yorker*, August 4, 2014, http://www.newyorker.com/magazine/2014/08/04/woman-2.

65 **violations of their rights:** For an overview of the ongoing history of the violations of the rights of transgender persons, see the Web site of the National Center for Transgender Equality, http://transequality.org.

66 **refers to her "clitoris":** See Novic, *Alice in Genderland*, 188, 229.

67 **"weird characterizations of us all":** Lynn Conway to Andrea James, Apr. 10, 2003, reproduced in Lynn Conway, "The Bailey Investigation: How It All Began with a Series of E-Mail Alerts," http://ai.eecs.umich.edu/people/conway/TS/Bailey/Investigation%20start-up/Investigation%20start-up.htm (accessed July 26, 2014) and quoted in Dreger, "Controversy," 384.

69 **man was quite femme:** See Bailey, *Man Who Would Be Queen*, preface.

69 **not particularly good-looking:** Ibid., 180.

69 **well suited for sex work:** Ibid., 185.

70 **"Kim still possessed a penis":** Ibid., 182.

71 **tone-*dumb*:** I learned the concept of literal tone dumbness from Stephen Fry, *Moab Is My Washpot* (New York: Random House, 1997).

71 **twin studies:** See, for example, Richard C. Pillard and J. Michael Bailey, "Human Sexual Orientation Has a Heritable Component," *Human Biology* 70, no. 2 (1998): 347–65.

73 **"[and those] who have not":** Andrea James, "Invective Against J. Michael Bailey's 'The Man Who Would Be Queen.'" Originally published as a page at www.tsroadmap.com in May 2003 and subsequently removed; complete copy obtained from files of J. Michael Bailey and discussed in Dreger, "Controversy," 368–69.

73 **to the Northwestern Rainbow Alliance:** E-mail from Alice Dreger to the Northwestern Rainbow Alliance, May 11, 2006; also discussed in Dreger, "Controversy," 369.

73 **"The Blog I Write in Fear":** Alice D. Dreger, "The Blog I Write in Fear," May 13, 2006, www.alicedreger.com/in_fear.html (accessed July 26, 2014); discussed in Dreger, "Controversy," 369.

74 **writing to Andrea James:** Alice Dreger to Andrea James, e-mail May 16, 2006; discussed in Dreger, "Controversy," 369.

74 **"precious womb turd":** Andrea James to Alice Dreger, e-mail June 1, 2006; discussed in Dreger, "Controversy," 369.

74 **"We'll chat in person soon":** Andrea James to Alice Dreger, e-mail May 27, 2006; discussed in Dreger, "Controversy," 369.

74 **university police:** See Dreger, "Controversy," 369.

74 **"Photos of Lynn":** Lynn Conway, "Photos of Lynn," http://ai.eecs.umich.edu/people/conway/Photos/Lynn-TN/LC-photos.html (accessed July 26, 2014).

75 **had been nominated:** This is discussed in the next chapter.

75 **formal charges made against Bailey:** Deirdre McCloskey to Alice Dreger, Jan. 22, 2007, as discussed in Dreger, "Controversy," 389.

76 **"He wanted what he wanted":** Deirdre N. McCloskey, *Crossing: A Memoir* (University of Chicago Press, 1999), 18–19.

76 **as it does on a natal woman:** In *Crossing: A Memoir,* McCloskey writes on p. 41, "Men do not get water in their eyes from a shower because the browridge makes it drip beyond their eyelashes. (Deirdre was delighted after her facial operations that she could no longer keep her eyes open under a shower.)"

77 **"*While I readily admit to my own autogynephilia*":** Andrea James to Anne Lawrence, e-mail Nov. 9, 1998, emphasis added; quoted and discussed in Dreger, "Controversy," 387–88.

CHAPTER 3: TANGLED WEBS

79 **formal complaints as posted:** I review in more detail the charges made against Bailey and analyze their merit in Alice D. Dreger, "The Controversy Surrounding *The Man Who Would Be Queen:* A Case History of the Politics of Science, Identity, and Sex in the Internet Age," *Archives of Sexual Behavior* 37 (2008): 366–421.

79 **Kieltyka had sought out Bailey:** Interview with Charlotte Anjelica Kieltyka, Aug. 16, 2006, revised transcript received Sept. 22, 2006; discussed in Dreger, "Controversy," 372.

79 **she had played a woman:** Interview with Charlotte Anjelica Kieltyka, Aug. 17, 2006, revised transcript received Sept. 22, 2006; interview with Charlotte Anjelica Kieltyka, Aug. 21, 2006, revised transcript received Sept. 27, 2006; discussed in Dreger, "Controversy," 372.

79 **shared the video:** Kieltyka also provided me a copy of this tape: Charlotte Anjelica Kieltyka, "Becoming Real: Chuck to Anjelica" (self-produced, 1999).

80 **rituals:** Interviews with Kieltyka, Aug. 21 and 22, 2006, revised transcripts received Sept. 27, 2006; discussed in Dreger, "Controversy," 372–74.

80 **"dress rehearsals":** Interview with Kieltyka, Aug. 17, 2006, revised transcript received Sept. 22, 2006.

80 **feminine foundation of herself:** Interview with Kieltyka, Aug. 21, 2006, revised transcript received Sept. 27, 2006; discussed in Dreger, "Controversy," 374.

80 **in his Human Sexuality class:** Interviews with Kieltyka, Aug. 16 and 17, 2006, revised transcripts received Sept. 22, 2006, and interview with J. Michael Bailey, Aug. 8, 2006, revised transcript received Aug. 8, 2006; discussed in Dreger, "Controversy," 373.

81 **even in the nude:** Interview with Kieltyka, Aug. 16, 2006, revised transcript received Sept. 22, 2006; discussed in Dreger, "Controversy," 407.

81 **including on local television:** Interview with Kieltyka, Aug. 16, 2006, revised transcript received Sept. 22, 2006, and J. Michael Bailey to Alice Dreger, e-mail interview, Jan. 17, 2007; discussed in Dreger, "Controversy," 407, 410. Kieltyka recorded and broadcast a presentation about her life on a local cable access channel, including a segment in which she is sitting in a television studio surrounded by recording equipment, wearing a white bikini, and drinking a cocktail, explaining to the camera that she's a transgender woman. For that broadcast, she had also chosen to share video of herself pretransition as a man; see Kieltyka, "Becoming Real."

81 **requests for sex reassignment surgery:** Interview with Kieltyka, Aug. 16, 2006, revised transcript received Sept. 22, 2006; discussed in Dreger, "Controversy," 372.

81 **Bailey thought:** Interview with J. Michael Bailey, Aug. 8, 2006, revised transcript received Aug. 8, 2006; discussed in Dreger, "Controversy," 372–73.

81 **Bailey's letters:** J. Michael Bailey to Alice Dreger, e-mail interviews, Oct. 2 and 3, 2006; as discussed in Dreger, "Controversy," 372–73.

82 **the newspaper article:** Maegan Gibson, "True Selves," *Focus, Daily Northwestern*, Feb. 24, 1999, 1, 5.

82 **human sexuality educational videos:** *Human Sexuality Videoworkshop*, 14 modules on CD-ROM, (Boston: Allyn & Bacon, 2004).

82 **Kieltyka *did* keep trying:** Interviews with Kieltyka, Aug. 21 and 22, 2006, revised transcripts received Sept. 27, 2006; discussed in Dreger, "Controversy," 376.

82 **pseudonym for the book:** Bailey to Dreger, Jan. 17, 2007; discussed in Dreger, "Controversy," 410.

83 **"Cher is a star":** J. Michael Bailey, *The Man Who Would Be Queen: The Science of Gender-Bending and Transsexualism* (Washington, DC: Joseph Henry Press, 2003), 212.

83 **"a sexual fantasy, she says":** Robin Wilson, "'Dr. Sex': A Human-Sexuality Expert Creates Controversy with a New Book on Gay Men and Transsexuals," *Chronicle of Higher Education*, June 20, 2003, 8.

84 **"doesn't want her last name used":** Ibid.

85 **"hanged by them":** Interview with Charlotte Anjelica Kieltyka, Sept. 19, 2006, revised transcript received Sept. 22, 2006; also quoted in Dreger, "Controversy," 388.

85 **"Anjelica, aka Cher":** Anjelica Kieltyka to J. Michael Bailey, May 16, 2003, as quoted in Dreger, "Controversy," 388.

85 **"field trips" to Chicago:** Lynn Conway, "An Investigative Report into the Publication of J. Michael Bailey's Book on Transsexualism by the National Academies," http://ai.eecs.umich.edu/people/conway/TS/LynnsReviewOfBaileysBook2.html (accessed July 27, 2014); discussed in Dreger, "Controversy," 389.

85 **"interviewing Bailey's research subjects":** Lynn Conway, "Timeline of the Unfolding Events in the Bailey Investigation," Jan. 6, 2010, http://ai.eecs.umich.edu/people/conway/TS/Bailey/Timeline/Timeline%20spreadsheet.htm (accessed July 27, 2014); version of Dec. 31, 2006, retrieved Jan. 22, 2007, discussed in Dreger, "Controversy," 389.

86 **Jim Marks:** Jim Marks to Alice Dreger, e-mail interview, July 22, 2006, as quoted in Dreger, "Controversy, 396.

86 **McCloskey told Marks:** Deirdre McCloskey to Jim Marks, personal communication, Feb. 3, 2004; reproduced at Lynn Conway, "The Gay and Lesbian 'Lambda Literary Foundation' Disses All Transsexual Women by Nominating Bailey's Book for a GLB'T' Literary Award!" http://ai.eecs.umich.edu/people/conway/TS/Bailey/Lambda%20Literary%20Foundation.html (accessed July 27, 2014), as quoted in Dreger, "Controversy," 411.

87 **Marks wasn't sure what to make of all this:** Marks to Dreger, July 22, 2006; quoted in Dreger, "Controversy," 396.

87 **Marks found:** Marks to Dreger, July 22, 2006; quoted in Dreger, "Controversy," 396–97.

88 **"future publication on this site":** Conway, "Gay and Lesbian 'Lambda Literary Foundation'"; quoted in Dreger, "Controversy," 397.

88 **According to Marks:** Marks to Dreger, July 22, 2006; quoted in Dreger, "Controversy," 397.

88 **Marks insisted was not true:** Ibid.

88 **that Bailey was autogynephilic:** Dreger, "Controversy," 398.

88 **Lawrence had been fully cleared:** Interview with Anne A. Lawrence, Aug. 8, 2006, revised transcript received Aug. 17, 2006; discussed in Dreger, "Controversy," 395.

88–89 **"makes a real human connection":** Andrea James to the faculty of the Northwestern University Psychology Department, Sept. 15, 2003; quoted in Dreger, "Controversy," 398.

89 **"consider moving":** Joan Linsenmeier to Alice Dreger, Aug. 17, 2006; quoted in Dreger, "Controversy," 397–98.

89 **lawyer told him to shut up:** See Dreger, "Controversy," 393, 404.

89 **Wilson had personally witnessed:** See Wilson, "'Dr. Sex.'"

90 **three terribly sober dispatches:** See Robin Wilson, "Transsexual 'Subjects' Complain About Professors' Research Methods," *Chronicle of Higher Education*, July 25, 2003, 10; Robin Wilson, "Northwestern U. Psychologist Accused of Having Sex with Research Subject," *Chronicle of Higher Education*, Dec. 19, 2003, 17; and Robin Wilson, "Northwestern U. Will Not Reveal Results of Investigation into Sex Researcher," *Chronicle of Higher Education*, Dec. 10, 2004, 10.

90 **Wilson's editor sent me back boilerplate:** Bill Horne to Alice Dreger, Aug. 15, 2006; quoted in Dreger, "Controversy," 394.
91 **Conway refused, as did Juanita:** My attempts to get Conway and Juanita to speak on the record are documented in Dreger, "Controversy."
91 **Bailey as a fall guy:** See Dreger, "Controversy," 386.
92 **putting a human face on autogynephilia:** I explain this conclusion further in Dreger, "Controversy," part 5, "The Merit of the Charges."
93 **by three trans women:** Lynn Conway, ed., "A Second Woman Files Research Misconduct Complaints Against Bailey," July 14, 2003, http://ai.eecs.umich.edu/people/conway/TS/Bailey/SecondComplaint.html (accessed July 27, 2014); Lynn Conway, ed., "A Third Woman Files Research Misconduct Complaints against Bailey," July 23, 2003, http://ai.eecs.umich.edu/people/conway/TS/Bailey/ThirdComplaint.html (accessed July 27, 2014); Lynn Conway, ed., "A Fourth Trans Woman Files a Formal Complaint with the Vice-President of Research of Northwestern University Regarding the Research Conduct of J. Michael Bailey," July 30, 2003, http://ai.eecs.umich.edu/people/conway/TS/Bailey/FourthWomansComplaint.html (accessed July 27, 2014).
93 ***not even in the book:*** See Dreger, "Controversy," 407.
93 **given him permission to do so:** Interview with Charlotte Anjelica Kieltyka, Sept. 19, 2006, revised transcript received Sept. 22, 2006; Lynn Conway, ed., "Documentation of a Formal Complaint About J. Michael Bailey's Sexual Exploitation of a Research Subject, and of Northwestern University's Apparent Decision to Not Investigate Such Egregious Misconduct," http://ai.eecs.umich.edu/people/conway/TS/Bailey/KeyDocuments/Misconduct-12-11-03.html. See also Dreger, "Controversy," 407.
94 **"most hurtful book of his":** Conway, "Documentation of a Formal Complaint," quoted in Dreger, "Controversy," 403.
94 **subsequent divorce:** See Dreger, "Controversy," 403.
94 **Illinois Department of Professional Regulation:** Lynn Conway, ed., "Evidence and Complaints Filed Against J. Michael Bailey for Practicing as a Clinical Psychologist Without a License, and Then Subsequently Publishing Confidential Clinical Case-History Information Without Permissions," Apr. 6, 2004, http://ai.eecs.umich.edu/people/conway/TS/Bailey/Clinical/ClinicalComplaint.html (accessed July 27, 2014). See also Dreger, "Controversy," 392.
95 **driven home the point:** Interview with J. Michael Bailey, Aug. 8, 2006, revised transcript received Aug. 8, 2006; Bailey to Dreger, Oct. 2 and 3, 2006. See also Dreger, "Controversy," 371–72, 410.
95 **relevant Illinois regulations:** *Illinois Compiled Statutes*, 225 ILCS 15/1, chap. 111, para. 5351. For further analysis and documentation, see Dreger, "Controversy," 411.
95 **broadcast the claim:** Charlotte Anjelica Kieltyka to C. Bradley Moore, July 3, 2003, "Anjelica Kieltyka Files a Formal Complaint . . ." (texts of three formal complaints to Northwestern University), http://ai.eecs.umich.edu/people/conway/TS/Anjelica/Complaint.html (accessed July 27, 2014). See also Conway, "A Second Woman Files"; Conway, "A Third Woman Files"; and Conway, "A fourth Trans Woman Files." See also Dreger, "Controversy," 400–402.
96 **IRB regulations:** On the history of IRBs, see Laura Stark, *Behind Closed Doors; IRBs and the Making of Ethical Research* (Chicago: University of Chicago Press,

2011) and see Zachary M. Schrag, *Ethical Imperialism: Institutional Review Boards and the Social Sciences, 1965–2009* (Baltimore: Johns Hopkins University Press, 2010).

96 **formalized scientific studies:** See Dreger, "Controversy," 377, 402.

98 **notarized affidavit:** Conway, "Documentation of a Formal Complaint"; also quoted in Dreger, "Controversy," 402–3.

98 **erotic semi-nude photo:** See Dreger, "Controversy," 371.

98 **rare public statement:** J. Michael Bailey, "Academic McCarthyism," *Northwestern Chronicle,* Oct. 9, 2005, http://archive.today/shRzY (accessed August 29, 2014); discussed in Dreger, "Controversy," 404.

98 **documentary evidence:** Provided by J. Michael Bailey to Alice Dreger, July 20, 2006 and discussed in Dreger, "Controversy," 404–5.

99 **Deb Bailey:** Interview with Deb Bailey, Aug. 9, 2006, no revisions to transcript; and Deb Bailey to Alice Dreger, e-mail interview, Jan. 7, 2007; also documented in Dreger, "Controversy," 404–5.

99 **"accountable for his actions":** See Conway, "Documentation of a Formal Complaint," quoted in Dreger, "Controversy," 403–4.

99 **He was adamant:** J. Michael Bailey to Alice Dreger, e-mail interview, July 19, 2006; see also Dreger, "Controversy," 404.

99 **Juanita wasn't interested in talking to me:** I document my attempts to speak to Juanita in Dreger, "Controversy," 371.

99 **"he couldn't get it up":** Interview with Kieltyka, Sept. 21, 2006, revised transcript received Sept. 27, 2006, quoted in Dreger, "Controversy," 405.

100 **"I wasn't enthusiastic":** Ibid.

100 **"Followed by narcissistic rage":** For elaboration of Lawrence's read on the matter, see Anne A. Lawrence, "Shame and Narcissistic Rage in Autogynephilic Transsexualism," *Archives of Sexual Behavior* 37, no. 3 (2008): 457–61.

101 **publishing narratives from trans women:** See Anne A. Lawrence, *Men Trapped in Men's Bodies: Narratives of Autogynephilic Transsexualism* (Doetinchem, Netherlands: Springer, 2013).

101 **Kiira Triea:** See Kiira Triea, "Power, Orgasm, and the Psychohormonal Research Unit," in *Intersex in the Age of Ethics,* ed. Alice Dreger (Hagerstown, MD: University Publishing Group, 1999): 141–44.

104 **Carey's piece was published:** Benedict Carey, "Criticism of a Gender Theory, and a Scientist Under Siege," *New York Times,* Aug. 21, 2007, www.nytimes.com/2007/08/21/health/psychology/21gender.html (accessed July 27, 2014).

105 **McCloskey wrote to the *New York Times*:** Deirdre McCloskey to *New York Times,* Aug. 24, 2007, www.deirdremccloskey.com/docs/times.pdf (accessed July 27, 2014).

105 **public radio in the Bay Area:** "Forum," KQED, Aug. 22, 2007; transcript at www.alicedreger.com/kqed_forum_transcript.html (accessed July 27, 2014). For a summary of incorrect statements by Roughgarden and my responses to them, see http://alicedreger.com/kqed_forum_corrections.html.

106 **Robin Mathy was filing ethics charges:** See footnote on p. 509 of Alice D. Dreger, "Response to the Commentaries on Dreger (2008)," *Archives of Sexual* Behavior 37 (2008): 503–10; see also Michael Gsovski, "Debate Resumes on Methods of Psych Professor's Research," *Daily Northwestern,* Mar. 18, 2008,

http://dailynorthwestern.com/2008/03/18/archive-manual/debate-resumes-on-methods-of-psych-professors-research (accessed July 27, 2014).

CHAPTER 4: A SHOW-ME STATE OF MIND

107 **chronic Lyme disease:** See David Grann, "Stalking Dr. Steere over Lyme Disease," *New York Times Magazine,* June 17, 2001, www.nytimes.com/2001/06/17/magazine/17LYMEDISEASE.html.

107 **fibromyalgia:** See Alex Berenson, "Drug Approved. Is Disease Real?" *New York Times,* Jan. 14, 2008, www.nytimes.com/2008/01/14/health/14pain.html.

107 **alleged childhood sexual abuse:** This refers to the story of Elizabeth Loftus, whose history I recount in Chapter 9. On her story, see Carol Tavris, "The High Cost of Skepticism," *Skeptical Inquirer* 26, no. 4 (July–Aug. 2002): 41–44, http://williamcalvin.com/2002/TavrisArticle.htm.

108 **alien abductions:** The book was Susan A. Clancy, *Abducted: How People Come to Believe They Were Kidnapped by Aliens* (Cambridge, MA: Harvard University Press, 2005).

109 **Ken Sher, my first interviewee:** Interview with Ken Sher, Oct. 30, 2008; Sher approved the passages about him on Dec. 3, 2012 (personal e-mail communication).

109 **exceedingly well documented:** See, e.g., the special issue, dedicated to the controversy, of *American Psychologist* 57, no. 3 (Mar. 2002), and Hollida Wakefield, "The Effects of Child Sexual Abuse: Truth Versus Political Correctness," *IPT Journal* 16 (2006), www.ipt-forensics.com/journal/volume16/j16_2.htm.

110 **"the Rind paper":** Bruce Rind, Philip Tromovitch, and Robert Bauserman, "A Meta-Analysis Examination of Assumed Properties of Child Sexual Abuse Using College Samples," *Psychological Bulletin* 124, no. 1 (July 1998): 22–53.

110 **Sher and Eisenberg had decided:** See Kenneth J. Sher and Nancy Eisenberg, "Publication of Rind et al. (1998): The Editors' Perspective," *American Psychologist* 57, no. 3 (Mar. 2002): 206–10.

111 **"or even altered":** Rind, Tromovitch, and Bauserman, "Meta-Analysis," 47.

111 **NAMbLA:** See NAMbLA, North American Man/Boy Love Association, "The Good News About Man/Boy Love" (1999), archived at https://web.archive.org/web/20140728213555/https://www.ipce.info/ipceweb/Documentation/Documents/99-112_nambla_statement.htm.

112 **virtual pitchmen for pedophilia:** See Sher and Eisenberg, "Publication."

112 **Tom Delay:** See Ellen Greenberg Garrison and Patricia Clem Kobor, "Weathering a Political Storm: A Contextual Perspective on a Psychological Research Controversy," *American Psychologist* 57, no. 3 (Mar. 2002): 165–75.

112 **A vote of 355 to 0:** See Garrison and Kobor, "Weathering," 172.

112 **"that Congress condemns":** House Concurrent Resolution 107 (106th Congress, 1999–2000): "Expressing the sense of Congress rejecting the conclusions of a recent article published by the American Psychological Association that suggests that sexual relationships between adults and children might be positive for children"; passed.

113 **"AND ARE IN ERROR":** Quoted in Sher and Eisenberg, "Publication," 206.

113 **APA kept Sher and Eisenberg apprised:** See Sher and Eisenberg, "Publication," 209.

113 **"on child welfare and protection issues":** See Garrison and Kobor, "Weathering," and Wakefield, "Effects."

114 **subverted in the service of politics:** See Scott O. Lilienfeld, "When Worlds Collide: Social Science, Politics, and the Rind et al. (1998) Child Sexual Abuse Meta-Analysis," *American Psychologist* 57, no. 3 (Mar. 2002): 176–88.

114 **"scientists in a professional field":** Quoted in Wakefield, "Effects."

114 **Dr. Laura on the radio:** This is also described in Sher and Eisenberg, "Publication," 209.

115 **"voting *for* pedophilia":** Sher and Eisenberg, "Publication," 206, n. 1, emphasis added.

115 **A *Natural History of Rape*:** Randy Thornhill and Craig T. Palmer, *A Natural History of Rape: Biological Bases of Sexual Coercion* (Cambridge, MA: MIT Press, 2000).

116 **Roughgarden had published:** Joan Roughgarden, review of *Evolution, Gender, and Rape* for *Ethology* 110, no. 1 (Jan. 2004): 76–78; quotation on p. 77.

116 **"excuse for criminal behavior":** Roughgarden, review, 76.

116 **Craig had told me in advance:** Our interview occurred on October 20, 2008, and Craig Palmer approved the passages about him on Oct. 26, 2010 (personal e-mail communication).

118 **in the *Sciences*:** Randy Thornhill and Craig T. Palmer, "Why Men Rape," *Sciences*, Jan.–Feb. 2000, 30–36.

118 **"Bill Clinton's behavior":** Interview with Craig Palmer, Oct. 30, 2008.

119 **Barbara Ehrenreich:** Barbara Ehrenreich, "How 'Natural' Is Rape?" *Time*, Jan. 31, 2000, 88.

119 **letter writer to the *Los Angeles Times*:** Doris C. Kagin, letter to the editor, *Los Angeles Times*, Mar. 13, 2000, E3.

119 ***Nashville Tennessean*'s:** Lawrence Spohn, "'Can't Help It' Theory Sparks Anger for Blaming Biology, Reproductive Instinct When a Man RAPES a Woman," *Nashville Tennessean*, Jan. 30, 2000.

119 ***Manchester Guardian*:** Michael Ellison, "The Men Can't Help It," *Manchester Guardian*, Jan. 25, 2000, 4.

119 ***Globe and Mail*:** "Are Men Natural-Born Rapists? Readers Weigh In," *Toronto Globe and Mail*, Feb. 12, 2000, D19.

119 **Susan Brownmiller:** See, for example, the interviews with Brownmiller in Janice D'Arcy, "Book Offers Radical Take on Rape," *Hartford Courant*, Feb. 6, 2000, A1, A8; and Martin Miller, "Rape," *Los Angeles Times*, Feb. 20, 2000, http://articles.latimes.com/2000/feb/20/news/cl-642.

119 **Brownmiller's highly influential opinion:** See Susan Brownmiller, *Against Our Will: Men, Women, and Rape* (New York: Martin Secker & Warburg, 1975).

120 **a pamphlet distributed:** Rape Prevention Education Program, "Resources Against Sexual Assault," University of California–Davis, Police Department, n.d.

123 **messages left on Randy's answering machine:** Interview with Randy Thornhill, Oct. 21, 2008, notes corrected and approved, Oct. 22, 2008.

124 **Elizabeth Eckstein:** Elizabeth Eckstein, "Rape: A Survivor's View," *Dallas Morning News*, Lifestyles, Feb. 2, 2000.

125 **interview with the *Boston Herald*:** Scripps Howard, "Study Says Rape Has Its Roots in Evolution," *Boston Herald*, Jan. 11, 2000, 3.

125 **from a guy serving time:** letter provided by Craig Palmer from his personal files.

127 **Joelle Ruby Ryan:** The exchange occurred on the WMST-L Listserv. Ryan's panel proposal was posted on Sept. 17, 2007; I responded on Sept. 19. The exchange continued and is available in the WMST-L archives at listserv .umd.edu.

127 **Conway functioned as Ryan's "mentor":** See Lynn Conway, "Report on Joelle Ruby Ryan's NWSA Panel Discussion published in the *Point Foundation's Mentoring Messenger,* Jan. 10, 2009, http://ai.eecs.umich.edu/people/conway/TS/News/US/NWSA/PF/Point_Foundation_Article_12-08.htm.

127 ***New York Times* coverage:** Benedict Carey, "Criticism of a Gender Theory, and a Scientist Under Siege," *New York Times,* Aug. 21, 2007, www.nytimes.com/2007/08/21/health/psychology/21gender.html.

128 **in my allotted fifteen minutes:** The paper I delivered was entitled "Activism in the Bailey Transsexualism Controversy Compared to Intersex Patients Rights Activism" and was presented on June 21, 2007.

128 **who was at her side but Juanita:** The page Lynn Conway mounted about the session included a photograph of Conway behind the video camera with Juanita sitting behind her: Conway, "Report."

128 **Panelists repeatedly defended:** Lynn Conway provides the papers and links to videos of the individual presentations; Lynn Conway, "Joelle Ruby Ryan Chairs NWSA Panel on Resisting Transphobia in Academia: The Event Alice Dreger Failed to Stop," June 27, 2008, http://ai.eecs.umich.edu/people/conway/TS/News/US/NWSA/NWSA_panel_on_resisting_transphobia_in_academia.html.

129 **interesting critiques of my work:** This was Katrina Rose; links to her paper and presentation, ibid.

130 **"Rosa Lee Klaneski":** The transcript of this text, taken from the video made, was provided to me by Rosa Lee Klaneski for this invited article: Alice Dreger, "In the Service of Galileo's Ghost: A Short Guide to History, Assault, and Ideology," in *History of Science Society Newsletter* 38, no. 4 (Oct. 2009). Rosa approved the content of that article on Aug. 19, 2009 (personal e-mail communication).

131 **"Alice, honey":** April Herndon and Rosa Lee Klaneski corroborated this account in personal e-mail communications of November 21, 2012.

134 **woman who had been Craig Palmer's dean:** I later learned from Craig Palmer that the dean who defended him was Elizabeth Grobsmith.

CHAPTER 5: THE ROT FROM WITHIN

141 ***Darkness in El Dorado*:** Patrick Tierney, *Darkness in El Dorado: How Scientists and Journalists Devastated the Amazon* (New York: Norton, 2000).

142 **Terence Turner and Leslie Sponsel:** See Terence Turner and Leslie Sponsel, letter to Louise Lamphere and Don Brenneis, "Re: Scandal About to Be Caused by Publication of Book by Patrick Tierney (*Darkness in El Dorado.* New York. Norton. Publication date: October 1, 2000)," http://anthroniche.com/darkness_documents/0055.htm. Many of the documents pertaining to the Tierney-Chagnon controversy are archived at the Anthropological Niche of Douglas W. Hume, http://anthroniche.com/darkness-in-el-dorado.html.

142 **hardly Turner and Sponsel's first attempt:** I discuss this in the article I published on this controversy; see Alice Dreger, "*Darkness*'s Descent on the

American Anthropological Association: A Cautionary Tale," *Human Nature* 22, no. 3 (2011): 225–46.

143 **The Guardian:** See Paul Brown, "Scientist 'Killed Amazon Indians to Test Race Theory,'" *The Guardian*, Sept. 23, 2000, www.theguardian.com/world/2000/sep/23/paulbrown.

143 **New Yorker article:** Patrick Tierney, "The Fierce Anthropologist," *New Yorker*, Oct. 9, 2000, 50–61.

143 **formal invitation to defend himself:** See Dreger, "*Darkness*'s Descent," 238–39.

143 **various other scholarly bodies:** Bruce Alberts, "Setting the Record Straight Regarding *Darkness in El Dorado*," Washington, DC, National Academy of Sciences, Nov. 9, 2000, http://anthroniche.com/darkness_documents/0538.htm; American Society of Human Genetics, "Response to Allegations Against James V. Neel in *Darkness in El Dorado*, by Patrick Tierney," *American Journal of Human Genetics* 70, no. 1 (Jan. 2002): 1–10; Max P. Baur, the IGES-ELSI Committee, et al., "Commentary on *Darkness in El Dorado* by Patrick Tierney," *Genetic Epidemiology* 21, no. 2 (Sept. 2001), 81–104; Society for Visual Anthropology, "Statement Approved by the Board of Directors and Unanimously Passed by the Membership of the Society for Visual Anthropology," Nov. 17, 2000, http://anthroniche.com/darkness_documents/0376.htm.

143 **University of Michigan:** Nancy Cantor, "Statement from University of Michigan Provost Nancy Cantor on the Book, *Darkness in El Dorado*, by Patrick Tierney," (Nov. 13, 2000), http://ns.umich.edu/Releases/2000/Nov00/r111300a.html.

143 **Susan Lindee:** Telephone interview with Susan Lindee, Dec. 12, 2008; approved revision received Dec. 15, 2008.

144 **issued an open letter:** See Lindee's letter to colleagues, Sept. 21, 2000, in Edward H. Hagen, Michael E. Price, and John Tooby, "Preliminary Report on *Darkness in El Dorado*," Department of Anthropology, University of California–Santa Barbara (unpublished, 2001), http://www.angelfire.com/sk2/title/ucsbpreliminaryreport.pdf, 61–62. In her letter, Lindee had also indicated she had found a telegram showing that Neel had obtained permission from the Venezuelan government to conduct vaccinations, but she later withdrew that claim after further review of the available evidence.

144 **Thomas Headland:** See remarks by Thomas Headland, open-microphone session, American Anthropological Association meeting, Nov. 16, 2000. See also Thomas N. Headland, "When Did the Measles Epidemic Begin Among the Yanomami?" *Anthropology News* 42, no. 1 (2001), 15–19, www.sil.org/~headlandt/measles1.htm.

144 **Diane Paul and John Beatty:** Diane Paul and John Beatty, "James Neel, *Darkness in El Dorado*, and *Eugenics: The Missing Context*," *Society for Latin American Anthropology* (electronic newsletter), no. 17 (Nov. 1, 2000), http://anthroniche.com/darkness_documents/0380.htm.

144 **portrayal of Neel as a Nazi-like eugenicist:** Ibid.; Susan Lindee, letter to American Anthropological Association, Nov. 16, 2000, read into the record, open-microphone session, American Anthropological Association meeting, Nov 16, 2000, retrieved from audio recordings and transcripts; Robert S. Cox, "Salting Slugs in the Intellectual Garden: James V. Neel and Scientific Controversy in the Information Age," *Mendel Newsletter*, Feb. 2001, www.amphilsoc.org/mendel/2001.htm#slugs.

144 **before Chagnon was even born:** For an example, see "Letter from Professor Jane Lancaster," in Hagen, Price, and Tooby, *Preliminary* Report, pp. 79–80.

145 **"swashbuckling misogynist":** Open-microphone session, American Anthropological Association, Nov 16, 2000, audio recordings and transcripts. The phrase "swashbuckling misogynist" comes from the remarks of William Vickers.

145 **spreading Ebola around Africa:** This claim was made by Omara Ben Abe in his remarks at the open-microphone session, ibid.

145 **Some anthropologists did try to fight back:** See John Tooby, "Jungle Fever," *Slate* (Oct. 25, 2000), http://www.slate.com/articles/news_and_politics/hey_wait_a_minute/2000/10/jungle_fever.html, and see Hagen, Price, and Tooby, *Preliminary Report.*

145 **launched a referendum:** American Anthropological Association, "Referendum on *Darkness in El Dorado* & Danger to Immunization Campaign," adopted Nov. 2003, http://www.aaanet.org/cmtes/ethics.Referendum-on-Darkness-in-El-Dorado-Task-Force.cfm?renderforprint=1.

145 **ratio of 11 to 1:** Approximately 14.5 percent of those eligible to vote on this AAA referendum did so: Kimberley Baker, AAA section & governance coordinator, to Alice Dreger, personal e-mail communication, Jan. 4, 2011; quoted in Dreger, "*Darkness*'s Descent," 229.

145 **another referendum:** American Anthropological Association, "Referendum #3: To Rescind the El Dorado Task Force Report" (2005), http://www.aaanet.org/stmts/05ref_eldorado.htm.

145 **Task Force Report:** American Anthropological Association, *El Dorado Task Force* papers, submitted to the Executive Board as a final report May 18, 2002, 2 vols. (Washington, D.C.: American Anthropological Association, 2002). As noted below, the Report was eventually removed from the AAA Web site but is still available (along with a treasure trove of related documents) at the AnthroNiche Web site of Douglas W. Hume. See http://anthroniche.com.darkness-in-el-dorado/archived-resources/position-statements.html.

145 **ratio of about 2.5 to 1:** Approximately 11 percent of those eligible to vote did so; Kimberley Baker, AAA section & governance coordinator, to Dreger, personal e-mail communication, Jan. 4, 2011;quoted in Dreger, "*Darkness*'s Descent," 229.

146 **to kill each other:** See the remarks by Davi Kopenawa in vol. 2, p. 25, of the Task Force Report.

146 **Chagnon's story:** Interview with Napoleon A. Chagnon, Traverse City, Michigan, Jan. 4–5, 2009; approved version returned Jan. 22, 2009. I asked Napoleon Chagnon to check all personal material about him in this and the next chapter not otherwise included in approved versions of interview notes, and he did so, confirming the material on Oct. 20, 2012.

147 **Chagnon's 1968 monograph:** Napoleon Chagnon, *Yanomamö: The Fierce People* (New York: Holt, Rinehart & Winston, 1968).

148 **South American anthropologists:** Interview with Napoleon Chagnon, Jan. 4–5, 2009; see also Dreger, "*Darkness*'s Descent," 227–28.

148 **Chagnon wrote to Neel:** Interview with Napoleon Chagnon, Jan. 4–5, 2009; the letter from Chagnon to Neel and Roche was dated Dec. 2, 1996 (copy provided by Napoleon Chagnon).

148 **YANOMAMA-1968-INSURANCE:** Lindee mentioned this folder in her Sept. 21, 2000, letter to colleagues, referenced above.

149 **he had essentially withdrawn the data:** See Raymond Hames, "The Political Uses of Ethnographic Description," in *Yanomami: The Fierce Controversy and What We Can Learn from It,* ed. Robert Borofsky (Los Angeles: University of California Press, 2005): 119–35.

150 **Ed Hagen, Michael Price, and John Tooby:** See Tooby, "Jungle Fever," and see Hagen, Price, and Tooby, *Preliminary Report.*

150 **when Wilson was presenting:** This story was recounted to me by Chagnon during our January 4-5, 2009, interview and also by Edward O. Wilson in our telephone interview on Aug. 24, 2009.

151 **AAA meeting in the 1970s:** Chagnon also tells this story in *Noble Savages: My Life Among Two Dangerous Tribes—the Yanomamö and the Anthropologists* (New York: Simon & Schuster, 2013), 384.

152 ***Margaret Mead and Samoa:*** Derek Freeman, *Margaret Mead and Samoa: The Making and Unmaking of an Anthropological Myth* (Cambridge, MA: Harvard University Press, 1983).

152 ***The Fateful Hoaxing of Margaret Mead:*** Derek Freeman, *The Fateful Hoaxing of Margaret Mead: A Historical Analysis of Her Samoan Research* (Boulder, CO: Westview Press, 1999).

153 **"in the groves of Academe":** Freeman as quoted in Paul Shankman, *The Trashing of Margaret Mead: Anatomy of an Anthropological Controversy* (Madison: University of Wisconsin Press, 2009), 10.

153 **fiction he had spun as nonfiction:** For a complete account of Freeman's mistreatment of Mead, see Shankman, *Trashing.*

154 **"collected throughout her fieldwork":** Martin Orans, *Not Even Wrong: Margaret Mead, Derek Freeman, and the Samoans* (Novato, CA: Chandler & Sharp, 1996), 99.

154 **"from the quicksand of controversy":** Shankman, *Trashing,* 19.

154 **"over 40% were sexually active":** Paul Shankman, "The 'Fateful Hoaxing' of Margaret Mead: A Cautionary Tale," *Current Anthropology* 54, no. 1 (Feb. 2013), 51–70; quotation from p. 59.

154 **"address important public issues":** Shankman, *Trashing,* 108.

155 **"critical junctures in his argument":** Ibid., 12.

155 **get the informant to turn on Mead:** Shankman, "'Fateful Hoaxing.'"

155 **"these two women in Mead's field notes":** Ibid., 59.

155 **even to himself:** Shankman, *Trashing,* 60–61.

155 **threatened those who did:** Ibid., 38.

155 **"was a Soviet agent":** Ibid., 54.

156 **"but we can't say it!":** Ibid., 56.

CHAPTER 6: HUMAN NATURES

157 **in anthropology or journalism:** See Alice Dreger, "*Darkness*'s Descent on the American Anthropological Association: A Cautionary Tale," *Human Nature* 22 (2011): 225–46.

157 ***The Highest Altar:*** Patrick Tierney, *The Highest Altar: The Story of Human Sacrifice* (New York: Viking, 1989).

157 **Chicago Public Radio:** Interview of Patrick Tierney by Victoria Lautman on WBEZ Chicago (Nov. 22, 2000), transcribed by Valerie Thonger.

158 **previous scholars who had looked:** See, e.g., Edward H. Hagen, Michael E. Price, and John Tooby, "Preliminary Report on *Darkness in El Dorado*," Department of Anthropology, University of California–Santa Barbara (unpublished, 2001), http://www.angelfire.com/sk2/title/ucsbpreliminaryreport.pdf.

158 **"named Marcel Roche":** Patrick Tierney, *Darkness in El Dorado: How Scientists and Journalists Devastated the Amazon* (New York: Norton, 2001), 60.

158 ***New Yorker* article:** Patrick Tierney, "The Fierce Anthropologist," *New Yorker*, Oct. 9, 2000, 50–61; see p. 57.

158 **article Chagnon had coauthored:** James V. Neel, Willard R. Centerwall, Napoleon A. Chagnon, and H. L. Casey, "Notes on the Effects of Measles and Measles Vaccine in a Virgin-Soil Population of South American Indians," *American Journal of Epidemiology* 91, no. 4 (1970): 418–29.

158 ***Yanomami Warfare*:** R. Brian Ferguson, *Yanomami Warfare: A Political History* (Sante Fe, NM: School of American Research Press, 1995).

159 **Ferguson told me:** Telephone interview with R. Brian Ferguson, July 28, 2009; corrections received Oct. 1 and 20, 2009.

159 **"important resource for my research":** Tierney, *Darkness*, xvii.

160 **confirmed in an e-mail:** Martins to Dreger, personal e-mail communication, June 5, 2009; quoted in Dreger, "*Darkness*'s Descent," 231.

161 **Chagnon had written to Hames:** Napoleon A. Chagnon to Raymond Hames, personal e-mail communication, Nov. 6, 1995; quoted with permission in Dreger, "*Darkness*'s Descent," 231.

161 **"appear to be *deliberately* fraudulent":** Hagen, Price, and Tooby, "*Preliminary Report*," 1.

162 **Turner was regularly making flight connections:** Terence Turner, telephone interview with Alice Dreger, Feb. 4, 2009; approved notes returned Feb. 8, 2009.

162 **Turner acknowledged to me:** Ibid.

162 **in part to go after Chagnon:** See, e.g., Lêda Leitão Martins, "On the Influence of Anthropological Work and Other Considerations on Ethics," *Public Anthropology: Engaging Ideas*, May 27, 2001, http://anthroniche.com/darkness_documents/0480.htm.

162 **Martins had publicly taken Chagnon to task:** For Martins's use of the truncated quotation, see Martins, "On the Influence." For a full translation of the quotation, which originally appeared in the Brazilian magazine *Veja*, see Robert Borofsky, ed., *Yanomami: The Fierce Controversy and What We Can Learn from It* (Los Angeles: University of California Press, 2005), 309.

163 **Salesian missionaries, with whom he had come to blows:** These disputes are discussed in Napoleon Chagnon, *Noble Savages: My Life Among Two Dangerous Tribes—the Yanomamö and the Anthropologists* (New York: Simon & Schuster, 2013).

164 **on handout tables at an AAA conference:** See Robin Fox, "Evil Wrought in the Name of Good," *Anthropology Newsletter* 35 (Mar. 1994): 2; Eric R. Wolf, "Demonization of Anthropologists in the Amazon," *Anthropology Newsletter* 35 (Mar. 1994), 2.

164 **distributed by the Salesians:** Frank A. Salamone, "Theoretical Reflections on the Chagnon-Salesian Controversy," in Frank A. Salamone and Walter R. Adams, eds., *Explorations in Anthropology and Theology* (Lanham, MD: University Press of America, 1997), 91–112.

164 **"Last Tribes of El Dorado":** Patrick Tierney, "*Last Tribes of El Dorado: The Gold Wars in the Amazon Rain Forest*" (scheduled for New York: Viking, 1994, apparently never published; page citations are from bound advance uncorrected proofs obtained via interlibrary loan).

164 **Viking wouldn't give:** I discuss this in Dreger, "*Darkness*'s Descent," 234.

165 **pass himself off as a Chilean gold miner:** Tierney, "*Last Tribes*," 29, 75, 87, 88, 131.

165 **carried mercury into the rain forest:** Ibid., 71.

165 **illegally purchased a shotgun:** Ibid, 71.

165 **without first undergoing appropriate quarantine:** Ibid., 172, 181.

165 **without first obtaining the required legal permission:** Ibid., 19, 124, 127. Tierney may have felt he was justified in doing this because he seems to have seen FUNAI as hopelessly corrupt; see pp. 182–83, 205, 210.

165 **self-confessed murderers:** Ibid., 69, 115, 138, 149, 163, 396.

165 **gotten another man killed:** Ibid., 327.

165 **housed, fed, protected, and encouraged by local Roman Catholic priests:** Ibid., 30, 50, 120, 216, 229, 231, 234, 272, 297, 298.

166 **Father Saffirio responded:** Interview by Alice Dreger of Giovanni Saffirio, Cleveland, July 8, 2009; approved notes received Aug. 12, 2009.

166 **"in Roraima":** Ibid.

167 **"big picture of a fine scholar":** Ibid.

167 **Frechione informed me:** Interview by Alice Dreger of John Frechione, Pittsburgh, July 8, 2009; approved notes returned July 30, 2009.

168 **2001 interview with Brandon Centerwall:** John Frechione interview of Brandon S. Centerwall, Oct. 27, 2001, http://anthroniche.com/darkness_documents/0102.htm.

168 **Turner had Brandon *on record*:** Regarding the additional supporting evidence from Terence Turner, see Dreger, "*Darkness*'s Descent," 232–33.

168 **suggesting that Humbert Humbert:** Brandon S. Centerwall, "Hiding in Plain Sight: Nabokov and Pedophilia," *Texas Studies in Literature and Language* 32, no. 3 (Fall 1990), 468–84.

168 **I wrote to ask him to confirm:** Alice Dreger to Brandon Centerwall, personal e-mail communication, Feb. 11, 2009.

168 **I wrote again five days later:** Alice Dreger to Brandon Centerwall, personal e-mail communication, Feb. 16, 2009.

168 **"or sharing it with others":** Brandon Centerwall to Alice Dreger, personal e-mail communication, Feb. 18, 2009.

169 **His four-page letter:** Brandon Centerwall to Alice Dreger, personal communication, Feb. 18, 2009, e-mail received Feb. 20, 2009.

171 **teaming up with Andrew Wakefield:** Interview by Dreger of Frechione.

172 **University of Pittsburgh:** I wrote to the University of Pittsburgh on July 7, 2009. A response came from Kathleen M. Dewalt on July 23, and I answered on July 27. On July 30, Dewalt wrote to say Tierney "is not currently appointed." I answered

on July 31: "Because your message of July 23 used the present tense for Patrick Tierney's appointment at the Center for Latin American Studies, I take it that the ending date of the appointment can be noted in my work as late July, 2009. . . . I assume also your message means Mr. Tierney no longer has any appointment with the University of Pittsburgh. If I have any of this incorrect, please let me know. If I do not hear from you further, I'll assume I have these facts right." Dewalt did not correct my understanding.

172 **Robert Cox:** Robert S. Cox, "Salting Slugs in the Intellectual Garden: James V. Neel and Scientific Controversy in the Information Age," *Mendel Newsletter,* Feb. 2001, www.amphilsoc.org/mendel/2001.htm#slugs.

172 **Charlie took me down to the stacks:** This visit occurred on June 30, 2009. Charles Greifenstein reviewed my draft description of this visit and in reply suggested no changes except perhaps mentioning more of the security aspects of the APS archive (Charles Greifenstein to Alice Dreger, personal e-mail communication, Jan. 26, 2011).

173 **James Neel to Mr. Hobert E. Lowrance:** James V. Neel to Hobert E. Lowrance, Mission Aviation Fellowship, Apr. 4, 1968, copy in Neel papers, American Philosophical Society, Philadelphia.

174 **Thomas Headland had confirmed:** See Thomas N. Headland, "When Did the Measles Epidemic Begin Among the Yanomami?" *Anthropology News* 42, no. 1 (2001), 15–19, www.sil.org/~headlandt/measles1.htm.

175 **Peacock Commission:** James Peacock, Janet Chernela, Linda Green, Ellen Gruenbaum, Philip Walker, Joe Watkins, and Linda Whiteford, "Report to Louise Lamphere, President of the American Anthropological Association, and the Executive Board of the American Anthropological Association: Recommendation for Investigation of *Darkness in El Dorado,*" known as the Peacock Report, Jan. 21, 2001, copy provided to me by Raymond Hames, now retrievable at http://anthroniche.com/darkness_documents/0612.pdf.

175 **When Hames resigned:** Raymond Hames, "My Resignation Letter" (from El Dorado Task Force), 2002, http://anthroniche.com/darkness_documents/0514.htm; Raymond Hames telephone interview with Alice Dreger, June 23, 2009; approved version returned July 6, 2009.

176 **They had so rushed it:** Peacock et al., "Peacock Report," 3: "In order to meet the deadline of January 22 for circulation of reports to the Executive Board, this report is submitted without explicit approval by all members of the Task Force of this final draft."

176 **Trudy Turner told me:** Trudy R. Turner, telephone interview with Alice Dreger, Aug. 24, 2009; approved version returned Sept. 16, 2009.

176 **Janet Chernela:** Janet Chernela, telephone interview with Alice Dreger, Aug. 10, 2009, approved notes returned Aug. 15, 2009.

176 **Yanomamö spokesperson who claimed:** This is discussed in Dreger, "*Darkness*'s Descent," 239.

176 **Jane Hill:** Jane Hill, telephone interview with Alice Dreger, July 15, 2009; approved version received July 16, 2009.

177 **"I don't remember the circumstances":** Ibid.

177 **batch of photocopies:** Obtained via e-mail from Sarah Hrdy, Nov. 6, 2009.

177 **gave me permission:** Hill provided permission via e-mail to me on Nov. 6, 2009.

177 **"Burn this message":** Jane Hill to Sarah Hrdy, personal e-mail communication, Apr. 15, 2002; used with permission. Also reproduced in Dreger, "*Darkness*'s Descent," 237.

178 **Louise Lamphere:** I note in Dreger, "*Darkness*'s Descent," 240, that "I asked Lamphere to confirm or deny this on the record, and she has not."

178 **"disagreed with their theoretical bent":** Francesca Bray to Alice Dreger, personal e-mail communication, Oct. 9, 2009, used with permission.

181 **HBES meeting:** Alice Dreger, "Darwin's Dangerous Critics: Evolutionary Biology and Identity Politics in the Internet Age," paper presented at annual meeting, Human Behavior and Evolution Society, California State University–Fullerton, May 30, 2009.

181 **At UCSB:** See Hagen, Price, and Tooby, *Preliminary Report*.

182 **University of Michigan:** See Nancy Cantor, "Statement from University of Michigan Provost Nancy Cantor on the Book, *Darkness in El Dorado*, by Patrick Tierney," Nov. 13, 2000, http://ns.umich.edu/Releases/2000/Nov00/r111300a.html.

182 **Chuck Roselli:** See John Schwartz, "Of Gay Sheep, Modern Science and Bad Publicity," *New York Times*, Jan. 25, 2007, www.nytimes.com/2007/01/25/science/25sheep.html.

182 ***twenty thousand e-mails:*** This account is based in part on interviews with Roselli and Newman: Charles Roselli, telephone interview with Alice Dreger, Nov. 5, 2008, approved notes received Nov. 8, 2008; Jim Newman, telephone interview with Alice Dreger, Oct. 23, 2008, approved notes received Nov. 8, 2008. Roselli and Newman also reviewed a draft of this section on Oct. 11, 2012, and agreed the representation is accurate.

182 **"defend researchers this way":** Interview with Newman.

183 **"'back to work'":** Ibid.

CHAPTER 7: RISKY BUSINESS

187 **promoting a high-risk drug regimen:** See Alice Dreger, Ellen K. Feder, and Anne Tamar-Mattis, "Prenatal Dexamethasone for Congenital Adrenal Hyperplasia: An Ethics Canary in the Modern Medical Mine," *Journal of Bioethical Inquiry* 9, no. 3 (2012): 277–94. For examples of New's clinic's promotion of the intervention, see "Prenatal Diagnosis and Treatment of Congenital Adrenal Hyperplasia," Maria New Children's Hormone Foundation, www.newchf.org/testing.php (accessed July 30, 2014). See also Elizabeth Kitzinger, "Prenatal Diagnosis & Treatment for Classical CAH," *CARES Foundation Newsletter* 2, no. 1 (Winter 2003): 15, www.caresfoundation.org/productcart/pc/news_letter/winter02-03_page_9.htm. See also the discussion of Maria New's 2001 presentation below.

187 **Dr. Maria New:** See "Biography: Dr. Maria Iandolo New," at http://www.nlm.nih.gov/changingthefaceofmedicine/physicians/biography_234.html.

187 **recommend the intervention:** See, for example, Kitzinger, "Prenatal Diagnosis," and the discussion of New's 2001 presentation below. The CARES (Congenital Adrenal Hyperplasia Research, Education & Support) Foundation has also enabled New's promotion of prenatal dexamethasone for CAH by, for example, posting New's biography calling hers "the only large center that provides prenatal diagnosis of CAH and prenatal treatment of affected females to

prevent genital ambiguity," at www.caresfoundation.org/productcart/pc/scientific_medical.html.

187 **"found safe for mother and child":** See "Prenatal Diagnosis and Treatment," Maria New Children's Hormone Foundation.

187 **studies of efficacy and long-term safety:** For a review of how little was actually known about the safety and efficacy of prenatal dexamethasone for CAH in 2010, see Mercè M. Fernández-Balsells et al. "Prenatal Dexamethasone Use for the Prevention of Virilization in Pregnancies at Risk for Classical Congenital Adrenal Hyperplasia Because of 21-Hydroxylase (CYP21A2) Deficiency: A Systematic Review and Meta-Analyses," *Clinical Endocrinology* 73, no. 4 (2010): 436–44. This article was published after the OHRP and FDA investigations (discussed in the next chapter) were completed, but the absence of data that it demonstrates was readily apparent to anyone who conducted a basic medical literature search.

187 **changing brain development:** See, for example, Hideo Uno et al. "Neurotoxicity of Glucocorticoids in the Primate Brain," *Hormones and Behavior* 28, no. 4 (Dec. 1994): 336–48. See also the concerns raised in Svetlana Lajic, Anna Nordenström, and Tatya Hirvikoski, "Long-Term Outcome of Prenatal Treatment of Congenital Adrenal Hyperplasia," in Christa E. Flück and Walter L. Miller, eds., *Disorders of the Human Adrenal Cortex* (Basel: Karger, 2008), 82–98. For a discussion of concerns from animal studies about prenatal dexamethasone increasing cardiovascular disease risk, see Svetlana Lajic, Anna Nordenström, and Tatya Hirvikoski, "Long-Term Outcome of Prenatal Dexamethasone Treatment of 21-Hydroxylase Deficiency," *Endocrine Development* 20 (2011): 96–105.

188 **"the only clinic in the United States":** See "Prenatal Diagnosis and Treatment," Maria New Children's Hormone Foundation; and Kitzinger, "Prenatal Diagnosis & Treatment for Classical CAH."

188 **funding to study, retrospectively:** Evidence of the outcomes study was readily available: "Determining the Long-Term Effects of Prenatal Dexamethasone Treatment in Children With 21-Hydroxylase Deficiency and Their Mothers," ClinicalTrials.gov, http://clinicaltrials.gov/ct2/show/NCT00617292.

188 **"*have not been determined*":** Ibid. (*Emphasis added*)

188 **large "accumulated" clinical population:** See p. 2 of M. I. New, "Androgen metabolism in childhood," grant application R01 HD00072-33A1 (approved), National Institutes of Health (New York: Cornell University Medical College, 1996) where New refers to the "large population of prenatally-treated infants" she had "accumulated" for study.

190 **trying to put a stop:** See Dreger, Feder, and Tamar-Mattis, "Prenatal Dexamethasone."

190 **a complete absence of any properly controlled scientific studies:** This was to be confirmed in the systematic review and meta-analysis published later that year by Fernández-Balsells et al., "Prenatal Dexamethasone." There has been no placebo-controlled trial of prenatal dexamethasone for CAH and no trials with outcomes judged by independent observers. While the Swedish team has performed a prospective study, New's U.S. group has tracked outcomes past birth only retrospectively, typically using low-level techniques like phone surveys and

questionnaires rather than physical examinations. For an example of the phone survey approach, see Maria I. New et al. "Extensive Personal Experience: Prenatal Diagnosis for Congenital Adrenal Hyperplasia in 532 Pregnancies," *Journal of Clinical Endocrinology & Metabolism* 86, no. 12 (2001): 5651–57. For an example of the use of retrospective questionnaires, see Heino F. Meyer-Bahlburg et al., "Cognitive and Motor Development of Children with and without Congenital Adrenal Hyperplasia after Early-Prenatal Dexamethasone," *Journal of Clinical Endocrinology & Metabolism* 89, no. 2 (2004): 610–14.

190 **see if the intervention works in animals:** For example, the FDA Web site assures consumers that, in the long process toward approval for drug indications, "companies, research institutions, and other organizations that take responsibility for developing a drug . . . must show the FDA results of preclinical testing in laboratory animals and what they propose to do for human testing." See Food and Drug Administration, "The FDA's Drug Review Process: Ensuring Drugs Are Safe and Effective," www.fda.gov/drugs/resourcesforyou/consumers/ucm143534.htm. In 1964, the Declaration of Helsinki codified as its first Basic Principle that clinical research "should be based on laboratory and animal experiments or other scientifically established facts." *Declaration of Helsinki*, 18th World Medical Assembly, Helsinki, 1964.

190 **The Swedish data:** See Tatya Hirvikoski, et al., "Cognitive Functions in Children at Risk for Congenital Adrenal Hyperplasia Treated Prenatally with Dexamethasone," *Journal of Clinical Endocrinology & Metabolism* 92 (2007): 542–48. It's interesting that an early report from New's group on cognitive outcomes hinted at similar possible adverse effects: see P. D. Trautman et al., "Effects of Early Prenatal Dexamethasone on the Cognitive and Behavioral Development of Young Children: Results of a Pilot Study," *Psychoneuroendocrinology* 20, no. 4 (1995): 439–49. See below for a discussion of later outcomes studies by New's group.

190 **The Swedish group had found:** See Svetlana Lajic et al., "Long-Term Somatic Follow-up of Prenatally Treated Children with Congenital Adrenal Hyperplasia," *Journal of Clinical Endocrinology & Metabolism* 83, no. 11 (1998): 3872–80.

191 **between 1 in 10,000 and 1 in 15,000:** Maguelone G. Forest, "Recent Advances in the Diagnosis and Management of Congenital Adrenal Hyperplasia due to 21-Hydroxylase Deficiency," *Human Reproduction Update* 10, no. 6 (Nov./Dec. 2004): 469–85. The NIH gives the frequency as "about 1 in 10,000 to 1 in 18,000" at *MedLine Plus*; see www.nlm.nih.gov/medlineplus/ency/article/000411.htm (accessed July 30, 2014).

191 **drew patients from around the world:** Chapter 9 includes a review of what numbers New reported to NIH.

192 **most common cause of congenital ambiguous genitalia:** National Institutes of Health, "Intersex," *MedLine Plus*, www.nlm.nih.gov/medlineplus/ency/article/001669.htm (accessed July 30, 2014). CAH actually comes in a number of different forms, and not all forms lead to masculinization in genetic females. The main type we're interested in here is the form called 21-hydroxylase deficiency.

192 **substantial natural variation:** See Medline Plus, "Congenital Adrenal Hyperplasia," at http://www.nlm.nih.gov/medlineplus/ency/article/000411.htm.

193 **ultimately identify as male:** Heino F. L. Meyer-Bahlburg, "What Causes Low Rates of Child-Bearing in Congenital Adrenal Hyperplasia?" *Journal of Clinical Endocrinology and Metabolism* 84, no. 6 (June 1999): 1844–47; Heino F. L. Meyer-Bahlburg et al., "Gender Development in Women with Congenital Adrenal Hyperplasia as a Function of Disorder Severity," *Archives of Sexual Behavior* 35, no. 6 (Dec. 2006): 667–84; and Arianne B. Dessens, Froukje M. E. Slijper, and Stenvert L. S. Drop, "Gender Dysphoria and Gender Change in Chromosomal Females with Congenital Adrenal Hyperplasia," *Archives of Sexual Behavior* 34, no. 4 (Aug. 2005): 389–97.

194 **opted to abort:** Selective abortion of females with CAH is reported, for example, in Arlene B. Mercado et al., "Extensive Personal Experience: Prenatal Treatment and Diagnosis of Congenital Adrenal Hyperplasia Owing to Steroid 21-Hydroxylase Deficiency," *Journal of Clinical Endocrinology and Metabolism* 80, no. 7 (July 1995): 2014–20.

194 **A 1984 paper:** Michel David and Maguelone G. Forest, "Prenatal Treatment of Congenital Adrenal Hyperplasia Resulting from 21-Hydroxylase Deficiency," *Journal of Pediatrics* 105, no. 5 (Nov. 1984): 799–803.

194 **making the intervention available:** See Chapter 9 for a review of the history of New's use of prenatal dexamethasone.

195 **Dr. New tells the families:** Maria I. New, lecture presented at conference for CARES Foundation, Weill Medical College of Cornell University, New York, Nov. 14, 2001.

196 **"done well in very few centers":** Ibid.

196 **giving birth prematurely:** Concerned researchers have been tracking possible unintended consequences of the use of prenatal steroids for premature birth risk; see, for example, National Institutes of Health Consensus Development Panel, "Antenatal Corticosteroids Revisited: Repeat Courses," Statement of NIH Consensus Development Conference, Aug. 17–18, 2000, *Obstetrics and Gynecology* 98, no. 1 (2001): 144–50; see also Noel P. French et al. "Repeated Antenatal Corticosteroids: Effects on Cerebral Palsy and Childhood Behavior," *American Journal of Obstetrics and Gynecology* 190, no. 3 (Mar. 2004): 588–95.

197 **families steadily learned:** For an example of a woman knowing to call Maria New, "a total stranger," as soon as she was pregnant with a fetus who might have CAH, see Catherine Elton, "A Prenatal Treatment Raises Questions of Medical Ethics," *Time*, June 18, 2010, http://content.time.com/time/health/article/0,8599,1996453,00.html.

197 **pregnancy category C:** FDA Pregnancy Categories, http://depts.washington.edu/druginfo/Formulary/Pregnancy.pdf (accessed July 30, 2014).

197 **"despite potential risks":** Ibid.

197 **the DES disaster:** For a history of DES, see Nancy Langston, *Toxic Bodies: Hormone Disruptors and the Legacy of DES* (New Haven, CT: Yale University Press, 2010).

197 **study published in 1953:** William J. Dieckmann et al., "Does the Administration of Diethylstilbestrol During Pregnancy Have Therapeutic Value?," *American Journal of Obstetrics and Gynecology* 66, no. 5 (Nov. 1953): 1062–82.

198 **serious question in 1971:** See Langston, *Toxic Bodies;* see also Centers for Disease Control, "About DES," www.cdc.gov/DES/CONSUMERS/about (accessed July 30, 2014).

198 **fatal vaginal cancer:** The first report of this cancer cluster's tie to DES was Arthur L. Herbst, Howard Ulfelder, and David C. Poskanzer, "Adenocarcinoma of the Vagina: Association of Maternal Stilbestrol Therapy with Tumor Appearance in Young Women," *New England Journal of Medicine* 284, no. 15 (Apr. 15, 1971): 878–81.

198 **Penny Stone:** See DES Action, "Meet the Woman Who Was the First to Connect DES and Cancer," *DES Action Voice*, no. 134 (Fall 2012): 8.

198 **reproductive cancers:** For a good overview of the harms caused by prenatal DES, see National Cancer Institute, "Diethylstilbestrol (DES) and Cancer," www.cancer.gov/cancertopics/factsheet/Risk/DES (accessed July 30, 2014).

198 **Frances Oldham Kelsey:** See Gardiner Harris, "The Public's Quiet Savior from Harmful Medicines," *The New York Times*, Sept. 13, 2010, http://www.nytimes.com/2010/09/14/health/14kelsey.html.

198 **ten thousand children in Europe:** Centers for Disease Control and Prevention, "Thalidomide," http://www.cdc.gov/healthcommunication/toolstemplates/entertainmented/tips/thalidomide.html.

199 **David Sandberg:** David Sandberg reviewed this account of our discussions and confirmed its accuracy, personal communication, Oct. 13, 2012.

200 **sixty to a hundred times:** Walter L. Miller, "Prenatal Treatment of Classic CAH with Dexamethasone: Con," *Endocrine News* (Apr. 2008): 16–18.

200 **All we had were reports:** The problems with this approach are alluded to in Fernández-Balsells et al., "Prenatal Dexamethasone."

200 **data coming out of Sweden:** See Hirvikoski et al., "Cognitive Functions."

201 **missing or choosing not to participate:** In 2010, New's chief collaborator, Heino Meyer-Bahlburg, admitted that "fewer than 50% of mothers and offspring have responded to questionnaires": in Phyllis W. Speiser et al., "A Summary of the Endocrine Society Clinical Practice Guidelines on Congenital Adrenal Hyperplasia due to Steroid 21-Hydroxylase Deficiency," *International Journal of Pediatric Endocrinology*, May 2010, www.ijpeonline.com/content/2010/1/494173; for quotation, see "3. Prenatal Treatment of CAH."

201 **"fetal programming":** The risk of fetal programming from prenatal dexamethasone was raised as early as 1997; see Jonathan R. Seckl and Walter L. Miller, "How Safe Is Long-Term Prenatal Glucocorticoid Treatment?," *Journal of the American Medical Association* 277, no. 13 (Apr. 2, 1997): 1077–79. For others raising the concern, see Hirvikoski et al., "Cognitive Functions." For more recent analyses of prenatal glucocorticoids and the programming of adult disease, see Anjanette Harris and Jonathan Seckl, "Glucocorticoids, Prenatal Stress, and the Programming of Disease," *Hormones and Behavior* 59, no. 3 (Mar. 2011): 279–89.

201 ***Wall Street Journal:*** Bernard Wysocki Jr., "As Universities Get Billions in Grants, Some See Abuses: Cornell Doctor Blows Whistle over Use of Federal Funds, Alleging Phantom Studies," *Wall Street Journal* (Aug. 16, 2005): A1.

202 **Sarafoglou went to the Feds:** Ibid.

202 **only about 80 percent of the time:** The authors of the formal pediatric endocrine consensus that emerged in 2010 (discussed in the next chapter) noted

the poor quality of efficacy data and concluded only that "the groups advocating and performing prenatal treatment appear to agree that it is effective in reducing and often eliminating virilization of female fetal genitalia and that the success rate is about 80–85%"; see Phyllis W. Speiser et al., "Congenital Adrenal Hyperplasia due to Steroid 21-Hydroxylase Deficiency: An Endocrine Society Clinical Practice Guideline," *Journal of Clinical Endocrinology and Metabolism* 95, no. 9 (Sept. 2010): 4133–60.

203 **had been warning:** Seckl and Miller, "How Safe."

203 **Miller had finally declared:** Miller, "Prenatal Treatment," 17.

203 **numerous medical societies:** See, for example, Joint LWPES/ESPE CAH Working Group, "Consensus Statement on 21-Hydroxylase Deficiency from the Lawson Wilkins Pediatric Endocrine Society and the European Society for Pædiatric Endocrinology," *Journal of Clinical Endocrinology and Metabolism* 87, no. 9 (Sept. 2002): 4048–53.

203 **American Academy of Pediatrics:** Jaime Frias, Lenore S. Levine, Sharon E. Oberfield, et al., "In Reply," (letter to the editor) *Pediatrics*, vol. 107, no. 4 (2001): 805. The letter referred to "the memory of the tragedies associated with prenatal use of dexamethasone and thalidomide," a line the authors later corrected to read "use of DES (diethylstilbestrol) and thalidomide"; see erratum, *Pediatrics* 107, no. 6 (June 2001): 1450, http://pediatrics.aappublications.org/content/107/6/1450.full.

204 **I wrote to her:** Alice Dreger to Maria New, e-mail communication, Dec. 8, 2009.

204 **Miami:** Maria I. New, "Long Range Outcome of Prenatal Treatment," conference presentation at 2nd World Conference, Hormonal and Genetic Basis of Sexual Differentiation Disorders and Hot Topics in Endocrinology, Jan. 15, 2010, Miami Beach.

205 **was very poor:** This was confirmed in Meyer-Bahlburg's report in Speiser et al., "A Summary of the Endocrine Society Clinical Practice Guidelines."

205 **at least one boy exposed:** The Swedish team formally published the evidence suggesting that boys exposed prenatally might be hypomasculinized in 2011 and 2012. See Lajic, Nordenström, and Hirvikoski, "Long-Term Outcome"; see also Tatya Hirvikoski et al., "Prenatal Dexamethasone Treatment of Children at Risk for Congenital Adrenal Hyperplasia: The Swedish Experience and Standpoint," *Journal of Clinical Endocrinology and Metabolism* 97, no. 6 (June 2012): 1881–83, doi:10.1210/jc.2012-1222.

205 **plugging prenatal dex:** See, for example, Maria I. New and Nathalie Josso, "Disorders of Sexual Differentiation," in Lee Goldman and J. Claude Bennett, eds., *Cecil Textbook of Medicine*, 21st ed. (Philadelphia: W. B. Saunders, 2000): 1297–1306; Maria I. New, Lucia Ghizzoni, and Karen Lin-Su, "An Update of Congenital Adrenal Hyperplasia," in Fima Lifshitz, ed., *Pediatric Endocrinology*, 5th ed. (New York: Informa Healthcare, 2007): 227–45.

205 **Marsha Rappley:** Conversation with Marsha Rappley; account confirmed by e-mail, Oct. 26, 2012.

206 **another e-mail message:** Alice Dreger to Maria New, e-mail communication, Jan. 24, 2010.

206 **from Jeffrey Silverstein:** Jeffrey H. Silverstein to Alice Dreger, e-mail communication, Jan. 26, 2010.

207 **I asked Dr. Silverstein to clarify:** Alice Dreger to Jeffrey H. Silverstein, e-mail communication, Jan. 26, 2010.

CHAPTER 8: DOCTOR, MY EYES

208 **American medical ethics regulations:** See the *Code of Federal Regulations*, Title 45, Part 46; for a sociological and historical analysis of institutional review boards (local ethics committees) in the United States, see Laura Stark, *Behind Closed Doors: IRBs and the Making of Ethical Research* (Chicago: University of Chicago Press, 2011).

208 **New consistently described:** For examples, see "Prenatal Diagnosis and Treatment of Congenital Adrenal Hyperplasia," Maria New Children's Hormone Foundation, www.newchf.org/testing.php (accessed July 30, 2014). See also Elizabeth Kitzinger, "Prenatal Diagnosis & Treatment for Classical CAH," *CARES Foundation Newsletter* 2, no. 1 (Winter 2003): 15, www.caresfoundation .org/productcart/pc/news_letter/winter02-03_page_9.htm. See also Maria I. New, lecture presented at conference for CARES Foundation, Weill Medical College of Cornell University, New York, Nov. 14, 2001.

208 **"to establish that prenatal treatment with dexamethasone is safe":** The same year New was describing dex as "safe" at the CARES Foundation meeting (ibid.), she told the NIH in her "Application for Continuation Grant," "We must now establish that prenatal treatment with dexamethasone is safe and has no long-term consequences"; Maria I. New, application for continuation grant, "Androgen Metabolism in Childhood," grant 5-R37-HD00072-37 (approved), National Institute of Child Health and Human Development (New York: Weill Cornell Medical College, 2001), quotation on p. 46. She specifically listed "prenatal treatment" as part of her "research plan" (p. 43). In the same grant application, New also told NIH: "This study is conducted . . . by FDA permission" (p. 47); I show below that there is no evidence she had such permission. As noted below, as late as 2006 she was still specifically naming dex-exposed fetuses as subjects of research for NIH grant purposes.

208 **"human subjects of research":** In her 2001 "Application for Continuation Grant" (ibid.), New described the sources of her human subjects this way: "Sources of human subjects are referrals from local and distant physicians who care for pregnant women at risk for having a fetus with CAH" (p. 47). Naming "the strengths of our group," New told the NIH, "We are the only group in the United States carrying out prenatal diagnosis and treatment of CAH and have thus accumulated a large population of prenatally treated patients to study" (p. 34).

209 **resources to weather criticism:** A classic example is that of Chester M. Southam who attempted infecting patients at the Jewish Chronic Disease Hospital with cancer, and who went on to be promoted within his field. See Chapter 17, "Illegal, Immoral, and Deplorable," in Rebecca Skloot, *The Immortal Life of Henrietta Lacks* (New York: Crown Publishers, 2010).

210 **code meant for Nazis:** See Susan M. Reverby, *Examining Tuskegee: The Infamous Syphilis Study and Its Legacy* (Chapel Hill: University of North Carolina Press, 2009), pp. 189, 193.

210 **"letter of concern":** Ellen K. Feder, Alice Dreger, Hilde Lindemann, et al., to the FDA Office of Pediatric Therapeutics, the Office for Human Research

Protections, Mount Sinai Medical Center, Weill Medical School of Cornell University, and Florida International University, February 3, 2010, known as "Letter of Concern from Bioethicists," reproduced at http://fetaldex.org/letter_bioethics.html.

211 "we agree with Dr. Miller": The quote from Miller appeared in Walter L. Miller, "Prenatal Treatment of Classic CAH with Dexamethasone: Con," *Endocrine News*, Apr. 2008, 16–18.

212 OHRP and the FDA had let us know: The OHRP response came from Kristina C. Borror to Ellen K. Feder and Anne Tamar-Mattis, Feb. 26, 2010. The FDA response came from Dianne Murphy to Ellen K. Feder, Feb. 8, 2010.

213 group of Boston clinicians: David A. Diamond et al., "Not Fetal Cosmetology," *Bioethics Forum*, Mar. 8, 2010, www.thehastingscenter.org/Bioethicsforum/Post.aspx?id=4528&blogid=140.

213 in the response: Alice Dreger, Ellen Feder, and Hilde Lindemann, "Prenatal Dex: Update and Omnibus Reply," *Bioethics Forum*, Mar. 18, 2010, www.thehastingscenter.org/Bioethicsforum/Post.aspx?id=4569&blogid=140.

214 "when the risks are non-trivial": Walter L. Miller to Alice Dreger, quoted with permission; also quoted on p. 284 of Alice Dreger, Ellen K. Feder, and Anne Tamar-Mattis, "Prenatal Dexamethasone for Congenital Adrenal Hyperplasia: An Ethics Canary in the Modern Medical Mine," *Journal of Bioethical Inquiry* 9 (2012): 277–94.

214 bioethics e-mail discussion list: This exchange occurred on the Medical College of Wisconsin bioethics Listserv (mcw-bioethics@mailman.mcw.edu) starting in late Jan. 2010.

214 report on 532 pregnancies: See Maria I. New et al., "Extensive Personal Experience: Prenatal Diagnosis for Congenital Adrenal Hyperplasia in 532 Pregnancies," *Journal of Clinical Endocrinology & Metabolism* 86, no. 2 (2001): 5651–57. An earlier paper from New and her team reporting on 239 pregnancies made no mention of any IRB approval or oversight; see Arlene B. Mercado et al., "Extensive Personal Experience: Prenatal Treatment and Diagnosis of Congenital Adrenal Hyperplasia Owing to Steroid 21-Hydroxylase Deficiency," *Journal of Clinical Endocrinology and Metabolism* 80, no. 7 (July 1995): 2014–20.

215 "preserve life or intellectual capacity": Phyllis W. Speiser et al., "Congenital Adrenal Hyperplasia Due to Steroid 21-Hydroxylase Deficiency: An Endocrine Society Clinical Practice Guideline," draft dated Aug. 31, 2009, 80 pp.; quotations at 13, 19. A different version (with the same conclusion) was eventually published as Phyllis W. Speiser et al., "Congenital Adrenal Hyperplasia due to Steroid 21-Hydroxylase Deficiency: An Endocrine Society Clinical Practice Guideline," *Journal of Clinical Endocrinology and Metabolism* 95, no. 9 (Sept. 2010): 4133–60. All of the lines quoted here remained the same from draft to final publication except for deletion of the line "the condition being treated, while fraught with emotional complexities, is directed toward a cosmetic outcome rather than aiming to preserve life or intellectual capacity." The line was replaced with: "Prenatal treatment of CAH is directed toward reducing the need for surgery, rather than toward preserving life or intellectual capacity."

215 "yield precise findings": Speiser et al., "Congenital Adrenal Hyperplasia," draft, 13.

215 the cosponsors: They were the American Academy of Pediatrics, Androgen Excess and PCOS Society, CARES Foundation, European Society for Endocrinology, European Society for Paediatric Endocrinology, Lawson Wilkins Pediatric Endocrine Society, and the Society of Pediatric Urology.

216 a formal call for responses: The target article abstract with a call for applications to respond was released by the *American Journal of Bioethics* on May 14, 2010. The *AJOB* target article was eventually published with responses (and with an amended title) as Laurence B. McCullough et al., "A Case Study in Unethical Transgressive Bioethics: 'Letter of Concern from Bioethicists' About the Prenatal Administration of Dexamethasone," *American Journal of Bioethics* 10, no. 9 (Sept. 2010): 35–45.

217 that 2002 position paper: See Joint LWPES/ESPE CAH Working Group, "Consensus Statement on 21-Hydroxylase Deficiency from the Lawson Wilkins Pediatric Endocrine Society and the European Society for Pædiatric Endocrinology," *Journal of Clinical Endocrinology and Metabolism* 87, no. 9 (Sept. 2002): 4048–53, at 4048.

218 over six hundred women: See "Prenatal Diagnosis and Treatment," Maria New Children's Hormone Foundation; and Kitzinger, "Prenatal Diagnosis & Treatment for Classical CAH."

218 "everybody else in the world put together": See New, presentation of Nov. 14, 2001.

219 lack of transparency: See my collaborator Hilde Lindemann's June 2011 resignation letter from *AJOB*'s editorial board at Brian Leiter's blog, http://leiterreports.typepad.com/blog/2011/06/editorial-misconduct-at-another-philosophy-journal-the-case-of-the-american-journal-of-bioethics.html. Glenn McGee responded to Lindemann in the comments.

219 "An Unethical Ethicist?": Brendan Borrell, "An Unethical Ethicist?," *Scientific American*, June 16, 2008, www.scientificamerican.com/article.cfm?id=glenn-mcgee. See also Brendan Borrell, "Alden March Bioethics Institute Picks Up the Pieces After Glenn McGee's Ouster," *Scientific American*, July 7, 2008, www.scientificamerican.com/article.cfm?id=bioethics-institute-picks.

220 I wrote to McCullough: This exchange occurred via e-mail on May 17 and 18, 2010. In an e-mail to me, copied to his coauthors and McGee, McCullough stated, "None of the authors of the paper: has any economic, professional, or any other kind of conflict of interest with regard to the content of our paper; has collaborated with Dr. New in her research, been funded on her grants, or served in any advisory capacity to her in her research; has ever published a paper with Dr. New; has ever written a prescription for a pregnant patient in one of Dr. New's trials; has ever 'acted as an ethics advisor to those administering, promoting, or researching this use of prenatal dex.'" As shown below, this was not true for McCullough and Chervenak.

221 *Journal of Urology*, in 2007: Jennifer Yang, Diane Felsen, and Dix P. Poppas, "Nerve Sparing Ventral Clitoroplasty: Analysis of Clitoral Sensitivity and Viability," *Journal of Urology* 178, no. 4, pt. 2 (Oct. 2007): 1598–1601.

222 "Bad Vibrations": Alice Dreger and Ellen K. Feder, "Bad Vibrations," *Bioethics Forum*, Hastings Center, June 16, 2010, www.thehastingscenter.org/Bioethicsforum/Post.aspx?id=4730.

222 Anne prepared legal letters: See Anne Tamar-Mattis to Jerry Menikoff, Director, Office for Human Research Protections, June 25, 2010, http://aiclegal.files.wordpress.com/2010/06/poppas-ohrp-letter.pdf.

222 Dan Savage pushed it hard for us: Dan Savage, "Female Genital Mutilation at Cornell University," *SLOG*, June 16, 2010, http://www.thestranger.com/slog/archives/2010/06/16/female-genital-mutilation-at-cornell-university&view=comments.

223 *Time* article: Catherine Elton, "A Prenatal Treatment Raises Questions of Medical Ethics," *Time*, June 18, 2010, http://content.time.com/time/health/article/0,8599,1996453,00.html (accessed July 30, 2014).

224 "if I wanted the treatment or not": Ibid.

224 *Endo Daily*: Anonymous, "Draft CAH Guideline Revealed Monday," *Endo Daily*, June 19–22, 2010, 8, www.nxtbook.com/tristar/endo/day4_2010/index.php?startid=8.

224 "This is not standard of care": Ibid. For the final version of the consensus document, see Speiser et al., "Congenital Adrenal Hyperplasia . . . Clinical Practice Guideline."

224 Meyer-Bahlburg: Heino F. L. Meyer-Bahlburg, "What Causes Low Rates of Child-Bearing in Congenital Adrenal Hyperplasia?" *Journal of Clinical Endocrinology and Metabolism* 84, no. 6 (June 1999): 1844–47; quotation at 1845–46.

225 "androgens on brain and behavior": Ibid., quotation at 1846.

225 *Annals of the New York Academy*: Saroj Nimkarn and Maria I. New, "Congenital Adrenal Hyperplasia due to 21-Hydroxylase Deficiency: A Paradigm for Prenatal Diagnosis and Treatment," *Annals of the New York Academy of Sciences*, no. 1192 (Apr. 2010): 5–11, quotation at 9.

225 "well-documented behavioral masculinization": Ibid, 9.

226 "not a reasonable goal of clinical care": Sandberg quoted in Elton, "Prenatal Treatment."

226 rates of tomboyism and lesbianism: Alice Dreger, Ellen K. Feder, and Anne Tamar-Mattis, "Preventing Homosexuality (and Uppity Women) in the Womb?" *Bioethics Forum*, Hastings Center, June 29, 2010, www.thehastingscenter.org/Bioethicsforum/Post.aspx?id=4754&blogid=140. We later expanded on this in Dreger, Feder, and Tamar-Mattis, "Prenatal Dexamethasone."

226 Dan Savage again helped us out: Dan Savage, "Doctor Treating Pregnant Women with Experimental Drug to Prevent Lesbianism," *SLOG*, June 30, 2010, http://www.thestranger.com/slog/archives/2010/06/29/doctor-treating-pregnant-women-with-experimental-drug-to-prevent-lesbianism.

226 "the anti-lesbian drug": See, for example, Sharon Begley, "The Anti-Lesbian Drug," *Newsweek*, July 2, 2010, www.newsweek.com/anti-lesbian-drug-74729 (accessed July 31, 2014).

226 the OHRP and the FDA indicated: Kristina C. Borror for OHRP to Ellen K. Feder and Alice Dreger, Sept. 2, 2010, reproduced at http://fetaldex.org/correspondence_files/OHRP_response_Sept_2_2010.pdf; Robert M. Nelson of FDA "through" Dianne Murphy of FDA to Kristina C. Borror for OHPR, Aug. 30, 2010, reproduced at http://fetaldex.org/correspondence_files/FDA_to_OHRP_Aug_30_2010.pdf.

227 **FDA investigator revealed:** Ibid.
227 **podcast with Larry McCullough:** Glenn McGee with Laurence McCullough, "A Case Study in Unethical Transgressive Bioethics," *Bioethics Channel*, Center for Practical Bioethics, Sept. 7, 2010, http://www.fluctu8.com/podcast-episode/a-case-study-in-unethical-transgressive-bioethics-84701-69055.html.
227 **Meyer-Bahlburg announced the Feds' nonfindings:** Heino F. L. Mayer-Bahlburg to SEXNET, Sexnet@Listserv.It.Northwestern.Edu, Sept. 3, 2010, "DSD Matter."

CHAPTER 9: DOOMED TO REPEAT?

230 **a hard time on thalidomide:** See Nancy Langston, *Toxic Bodies: Hormone Disruptors and the Legacy of DES* (New Haven: Yale University Press, 2010).
231 ***New York Review of Books:*** Elizabeth Allen et al., "Against 'Sociobiology,'" *New York Review of Books*, Nov. 13, 1975, http://www.nybooks.com/articles/archives/1975/nov/13/against-sociobiology.
231 **"from these monstrous crimes":** Quoted in Ullica Segerstrale, *Defenders of the Truth: The Sociobiology Debate* (Oxford, UK: Oxford University Press, 2000), 181.
231 **talking to Wilson:** Alice Dreger, telephone interview with Edward O. Wilson, Aug. 24, 2009.
233 **Elizabeth Loftus:** Elizabeth Loftus reviewed and agreed with this account of our conversation on February 13, 2012.
234 **real name is Nicole Taus:** See Carol Tavris, "Whatever Happened to 'Jane Doe'?" *Skeptical Inquirer* 32, no. 1 (Jan.–Feb. 2008), www.csicop.org/si/show/whatever_happened_to_jane_doe.
234 **saying her privacy was being violated:** See Carol Tavris, "The High Cost of Skepticism," *Skeptical Inquirer* 26, no. 4 (July–Aug. 2002): 41–44, http://www.csicop.org/si/show/high_cost_of_skepticism/.
234 **the two went on to publish:** Elizabeth F. Loftus and Melvin J. Guyer, "Who Abused Jane Doe? The Hazards of the Single Case History: Part I," *Skeptical Inquirer* 26, no. 3 (May–June 2002): 24–32, http://faculty.washington.edu/eloftus/Articles/JaneDoe.htm.
234 **In the end, they prevailed:** See Tavris, "Whatever Happened to 'Jane Doe'?"
235 **I hired a lawyer and sued:** *Alice Dreger v. U.S. Department of Health & Human Services, Food & Drug Administration, and Office for Human Research Protections*, United States District Court, Western District of Michigan, Southern Division, filed Oct. 3, 2011, 1:2011-cv-01059.
236 **ethics canary in the modern medical mine:** Hence the subtitle I gave our paper on the matter: Alice Dreger, Ellen K. Feder, and Anne Tamar-Mattis, "Prenatal Dexamethasone for Congenital Adrenal Hyperplasia: An Ethics Canary in the Modern Medical Mine," *Journal of Bioethical Inquiry* 9 (2012): 277–94.
236 **Cornell clinic in 1986:** This date comes from a faxed letter from Maria I. New to Jeff Cohen of Cornell's medical school dated August 19, 2004, obtained via FOIA, and is confirmed in Arlene B. Mercado et al., "Extensive Personal Experience: Prenatal Treatment and Diagnosis of Congenital Adrenal Hyperplasia Owing to Steroid 21-Hydroxylase Deficiency," *Journal of Clinical Endocrinology and Metabolism* 80, no. 7 (July 1995): 2014–20.

237 drug trials were not rigorous: See the critical analysis of New's studies in Mercè M. Fernández-Balsells et al., "Prenatal Dexamethasone Use for the Prevention of Virilization in Pregnancies at Risk for Classical Congenital Adrenal Hyperplasia Because of 21-Hydroxylase (CYP21A2) Deficiency: A Systematic Review and Meta-Analyses," *Clinical Endocrinology* 73, no. 4 (2010): 436–44.

237 some results: Mercado et al., "Extensive Personal Experience."

237 New's first Cornell IRB application: Maria I. New, IRB application (approved) for project entitled "Steroid 21-Hydroxylase Deficiency: Inborn error of steroid synthesis" (New York: New York Hospital-Cornell Medical Center Institutional Review Board: 1985), obtained via FOIA.

237 *sixteen years:* New noted that she was now checking the boxes in a letter to Owen Davis, Chair of the Cornell IRB, on March 21, 2001, subject "Annual Review Report for Protocol #0296-223CRC" (obtained via FOIA). That said, a year later, there is an IRB application from New for prenatal dex marked "received Mar 15 2002" on which she again did not check the boxes for "pregnant women" or "fetuses"; see Maria I. New, "Request for Approval of Investigation Involving Use of Human Subjects," Protocol 0296-223 (obtained via FOIA). Similarly, in 2003, the Cornell IRB approved a minor revision of New's IRB protocol to "clarify" that "The correct age criteria should read, 'Newborn to 100 years'" without apparently noticing that it should really have read "*fetuses* to 100 years." The request was from Maria I. New to David Behrman, Chair of the Cornell IRB, March 4, 2003, subject "Protocol #0296-223"; approval returned April 21, 2003 from Behrman to New (obtained via FOIA).

238 1985 consent form: This formed part of the 1985 IRB application noted below. We also discuss this in Dreger, Feder, and Tamar-Mattis, "Prenatal Dexamethasone."

238 "transient and reversible suppression": Ibid., p. 9e.

238 updated information: For example, New's consent forms did not incorporate notice of the potential harms in terms of temperament and behavior as reported by New's own group in P. D. Trautman et al., "Effects of Early Prenatal Dexamethasone on the Cognitive and Behavioral Development of Young Children: Results of a Pilot Study," *Psychoneuroendocrinology* 20, no. 4 (1995): 439–49, nor of the potential somatic effects (including failure to thrive and delayed psychomotor development) reported in Svetlana Lajic et al., "Long-Term Somatic Follow-up of Prenatally Treated Children with Congenital Adrenal Hyperplasia," *Journal of Clinical Endocrinology & Metabolism* 83, no. 11 (1998): 3872–80. Maternal "side" effects are also downplayed on New's consent forms in spite of reports in the literature of "significant maternal side effects" including "Cushingoid facial features, severe striae resulting in permanent scarring, and hyperglycemic response" in addition to "hypertension, gastrointestinal intolerance, or extreme irritability"; see S. Pang, A. T. Clark, L. C. Freeman, et al., "Maternal Side Effects of Prenatal Dexamethasone Therapy for Fetal Congenital Adrenal Hyperplasia," *Journal of Clinical Endocrinology & Metabolism*, vol. 75, no. 1 (1992): 249–53.

238 As late as 2004, Cornell's IRB: Maria I. New, Consent form for clinical investigation (IRB approved) for project entitled "Hypo- and hyperadrenal states/prenatal diagnosis and therapy" (New York: New York Presbyterian Hospital-Weill Medical College of Cornell University: 2004), obtained via FOIA.

238 **than any other researcher:** See, e.g., p. 2 of New's 1996 grant application, Maria I. New, "Androgen metabolism in childhood," grant application R01 HD00072-33A1 (approved), National Institutes of Health (New York: Cornell University Medical College, 1996); see also p. 34 of her 2001 application for continuation grant, Maria I. New, application for continuation grant, "Androgen Metabolism in Childhood," grant 5-R37-HD00072-37 (approved), National Institute of Child Health and Human Development (New York: Weill Cornell Medical College, 2001).

238 **2001 grant renewal application:** See New, 2001 "Application for Continuation Grant," 34; emphasis added.

239 **2,144:** See New, 2001 "Application for Continuation Grant," 42.

239 **By 1996, the NIH was specifically:** New's 1996 NIH application reported that "genital abnormalities and often multiple corrective surgeries needed affect social interaction, self image, romantic and sexual life, and fertility. As a consequence, many of these patients, and the majority of women with the salt-losing variant [of CAH], appear to remain childless and single. Preventative prenatal exposure is expected to improve this situation"; see p. 38 of Maria I. New, "Androgen metabolism in childhood," grant application R01 HD00072-33 (approved), National Institutes of Health (New York: Cornell University Medical College, 1996). In a related application packet, New specifically promised to try to determine "the success of DEX in suppressing behavioral masculinization"; see p. 17 of Maria I. New, "Androgen metabolism in childhood," grant application R01 HD00072-33A1 (approved), National Institutes of Health (New York: Cornell University Medical College, 1996). In her 2001 "Application for Continuation Grant," New reiterated the same interest in prenatal dex: "Our studies of the outcome in CAH patients with respect to gender, cognition, and social function will provide vital information on the validity of our [prenatal] treatment protocol"; quotation at 44.

239 **Kyriakie Sarafoglou was hired:** Anonymous, "Medical College Pays $4.4 Mil Settlement," *The Cornell Daily Sun*, Sept. 15, 2005, http://cornellsun.com/blog/2005/09/19/medical-college-pays-44-mil-settlement/. She was hired into this position in August of 2001 according to the report of an investigation conducted by Adam Asch of Cornell at the request of David Hajjar, Dean at Cornell's medical school, on November 4, 2002 (obtained by FOIA).

240 **financial and ethical irregularities:** See Bernard Wysocki Jr., "As Universities Get Billions in Grants, Some See Abuses: Cornell Doctor Blows Whistle over Use of Federal Funds, Alleging Phantom Studies," *Wall Street Journal*, Aug. 16, 2005): A1.

240 **concerning irregularities:** in a memo with the subject line "Protocol # ????-???," Sarafoglou noted to Cornell administrator Valerie Johnson that two of New's protocols appeared to be exact duplicates, "each requesting $172,025" in terms of budget." She asked, "How is this possible if they are separate studies?" and asked if the "same [100] patients will now undergo the same tests as they did" in another protocol; "Does this make sense?" New responded to Johnson in a letter dated December 5, 2002, that "I do not understand why the reviewers need to

continually compare two protocols. Further, there's no contraindication by the NIH to overlapping protocols." (Obtained via FOIA.)

240 **internal Cornell report:** See Asch to Hajjar, November 4, 2002. According to this report, Sarafoglou put her concerns to Gerald Loughlin, Chair of Pediatrics, on September 10, 2002.

240 **anguished message:** Kyriakie Sarafoglou to Kristina Borror of OHRP, December 28, 2003 (obtained via FOIA).

240 **"informed consent could be found":** Review of Neil H. White for OHRP, May 20, 2004 (obtained via FOIA).

240 **"children as subjects of research":** Review of Bruce Gordon for OHRP, May 21, 2004 (obtained via FOIA).

240 **angry memo:** This was obtained via FOIA and has a subject line referring to New's main CAH IRB protocol at Cornell (#0296-223). This memo is further discussed below.

241 **New reporting to her IRB:** Two examples located via FOIA: (1) Reviewer "Dr. Aledo" reporting on a meeting with Maria New on April 9, 2001, about her umbrella CAH IRB protocol: "112 accruals[,] 0 refusals[,] withdrawals[,] complaints." Aledo recommended approval for another year. (2) Reviewer "Dr. Aledo" reporting on a meeting with Maria New on April 8, 2002, about the same protocol: "35 accruals[,] 0 withdrawals[,] refusals[,] complaints." Aledo recommended approval for another year. See also the 1997 letter cited next.

241 **in a letter to her IRB:** Maria I. New to Dorothy Hilpmann, IRB Chair, April 4, 1997, subject "Annual Renewal Report for Protocol #0296-223CRC." (Obtained via FOIA.)

241 **a publication she had co-authored:** the publication cited in the letter was Mercado et al., "Extensive Personal Experience."

241 **undertook a massive review:** For a discussion of this earlier OHRP investigation, see Dreger, Feder, and Tamar-Mattis, "Prenatal Dexamethasone," pp. 289–90.

241 **extraordinary step of requiring review:** See Patrick J. McNeilly for OHRP to Antonio M. Grotto and Jeffrey M. Cohen of Weill College of Medicine of Cornell University, July 21, 2004, subject "Human research subject protections under Multiple Project Assurance (MPA) M-1185 and Federalwide Assurance (FWA) 93." (Obtained via FOIA.)

241 **letter dated May 24, 2004:** Patrick J. McNeilly for OHRP to Antonio M. Gotto, and Jeffrey M. Cohen of Weill College of Medicine of Cornell University, May 24, 2004; quotation on p. 7, item 15 (obtained via FOIA).

241 **a lot of back-and-forth:** See Jeffrey M. Cohen, Weill Medical College of Cornell, to Patrick J. McNeilly, OHRP, June 29 and August 31 2004. (Obtained via FOIA.)

242 **New wrote back a curt memo:** Maria I. New to Jeff Cohen of Cornell, fax transmission dated August 9, 2014. (Obtained via FOIA.)

242 **in practice:** Indeed, one interesting line in her grant renewal from 2001 seems to confirm this approach of treating the pregnant women as patients at Cornell while naming them as research subjects to NIH: "Sources of human subjects are referrals from local and distant physicians who care for pregnant women at risk for having a fetus with CAH." But a few lines later, New adds: "Prenatal diagnosis and treatment are performed for clinical indications and are not primarily

research purposes." See 2001 "Application for Continuation Grant," 47. See also the discussion of the fetuses treated at Mount Sinai, below.

242 **OHRP would later tell us:** Kristina C. Borror for OHRP to Ellen K. Feder and Alice Dreger, Sept. 2, 2010, reproduced at http://fetaldex.org/correspondence_files/OHRP_response_Sept_2_2010.pdf.

242 **Cornell's word to OHRP:** Mary Simmerling for Cornell to Kristina Borror for OHRP, July 29, 2010. (Obtained via FOIA.)

243 **until Sarafoglou's complaints:** The first audit I can find occurred in late 2002 as a response to Sarafoglou's complaints; see Asch to Hajjar, November 4, 2002. It only included 50 patient charts and the method of chart selection is unclear.

243 **angry whistleblowing memo:** As noted above, this was obtained via FOIA and has a subject line specifically referring to New's main CAH IRB protocol at Cornell (#0296-223).

243 **consistently led the NIH to believe:** All of Maria New's NIH grant materials from 1996 forward that I have obtained include discussions of prenatal dexamethasone treatments as part of her experimental research. Even when she discussed retrospective follow-up studies, she also specifically discussed new pregnancy exposures as part of her ongoing research plan. See below for a discussion of how this did not change when she moved to Mount Sinai.

244 **in 2006, in a research progress report:** Maria I. New, "Androgen metabolism in childhood," grant progress report 5-R37-HD00072-42 to Department of Health and Human Services, Public Health Services (New York: Mount Sinai School of Medicine: 2006), quotation on p. 3. (Obtained via FOIA.)

244 **letter from the head of Mount Sinai's IRB:** Jeffrey H. Silverstein, IRB Chair of Mount Sinai School of Medicine, to NIH, subject "GCO Project #04-0469 0001 01 PE," September 2, 2004 (obtained via FOIA). It is unlikely that the 2004 project called "prenatal diagnosis and treatment" was a retrospective study of the sort that Silverstein said in 2010 had IRB approval, because in his letter to the OHRP in 2010, Silverstein said that the title of the retrospective study was "Long Term Outcome in Offspring and Mothers of Dexamethasone-Treated Pregnancies at Risk for Classical Adrenal Hyperplasia Owing to 21-Hydroxylase Deficiency"; see Jeffrey H. Silverstein for Mount Sinai School of Medicine to Kristina C. Borror of OHRP, August 2, 2010 (obtained via FOIA). New's IRB applications at Cornell consistently distinguished between the prenatal interventions and the retrospective studies.

244 **same administrator who told the OHRP:** Silverstein to Borror, August 2, 2010.

244 **Mount Sinai administrator told OHRP:** Ibid., 8.

244 **New had written to the NIH:** Maria I. New to Duane Alexander, dated February 12, 2003, "Re: 4-R37HD00072-38." (Obtained via FOIA.)

245 **writing her a big check:** NIH Staff, administrative increase/administrative supplement, staff recommendation "re. Maria I. New," signed April 3, 4, and 7, 2003. (Obtained via FOIA.) In an email exchange on April 1, 2003, Barbara L. Pifel of Cornell told Angelos Bacas of NIH grant management that September 1, 2002 was "when Dr. New stopped receiving salary." (Obtained via FOIA.)

245 **perpetual motion machine:** Aron C. Sousa, "The Dex Diaries, Part 4: A Perpetual Motion Machine of NIH Funding?" Aug. 21, 2012, http://fetaldex.org/diary04.html.

246 **"which are part of the research":** Remark by "Dr. Aledo," minutes of the Committee on Human Rights in Research of Cornell University Weill Medical College, September 29, 2003. (Obtained via FOIA.)
246 **McCullough told OHRP:** Laurence B. McCullough to Kristina C. Borror of OHRP, April 23, 2010. (Obtained via FOIA.)
247 **at the medical schools of Cornell and Mount Sinai:** See Alice Dreger and Ellen K. Feder, "FDA Ethicist's Undisclosed Conflicts of Interest in Prenatal Dex Case," Apr. 17, 2014, http://impactethics.ca/2014/04/17/. fda-ethicists-undisclosed-conflicts-of-interest-in-prenatal-dex-case.
247 **served as "key personnel":** Frank A. Chervenak to Maria I. New, Apr. 15, 2003, regarding Rare Diseases Clinical Research Network grant proposal RR-03-008, reproduced at http://www.fetaldex.org/AJOB_Chervenak.html. (Obtained via FOIA.)
247 **Robert "Skip" Nelson:** See Robert M. Nelson of FDA "through" Diane Murphy of FDA to Kristina Borror for OHPR, Aug. 30, 2010.
247 **negotiating a new *AJOB* journal editorship in chief:** See Dreger and Feder, "FDA Ethicist's Undisclosed Conflicts."
247 **to hire an editorial assistant:** Vince Tolino, FDA Director of Ethics and Integrity, advised Nelson in e-mail discussions about this new position that it would look "cleaner" if a university used the money to pay a grad student to be an editorial assistant to Nelson. This correspondence is reproduced at http://fetaldex.org/updates_files/FOIA%20Nelson%20FDA%20AJOB-PR%20position.pdf.
247 **was keeping track:** Robert "Skip" Nelson to Jerry Menikoff, Kristina C. Borror, and Michael A. Carome, e-mail communication, Sept. 1, 2010, subject line "FDA Memo to OHRP re Dex for CAH" (obtained via FOIA): "Let us know when it will be posted."
247 **Nelson had told everybody:** See Nelson through Murphy to Borror.
248 **the 1996 exemption letter:** Solomon Sobel for the FDA to Maria I. New, Cornell Medical Center, February 7, 1996 (obtained via FOIA).
248 **without full FDA review:** In fact the 1996 letter (ibid.) notes, in boilerplate language, that an IND exemption can only be provided if "the route of administration, dosage level, patient population, and other factors do not significantly increase the risks," which surely would not have been true when aiming the intervention at fetuses.
248 **no recollection of the matter:** e-mail exchange between Alice Dreger and Solomon Sobel of the FDA, July 19, 2010. The reason I contacted Sobel at this time was that New had told a journal editor with whom I was communicating that "Prenatal dexamethasone treatment has been FDA approved by Dr. Sobel," and I was trying to understand what she meant. For details, see Dreger, Feder, and Tamar-Mattis, "Prenatal Dexamethasone."
248 **shredded during an FDA move:** Robert "Skip" Nelson to Jerry Menikoff, e-mail communication, June 14, 2010 (obtained via FOIA): "Thanks. I just learned that exemption letter documentation ([one inch of text redacted], certainly ones from 1996) were shredded for the move to White Oak [redacted]." See also Nelson's e-mail to Diane Murphy at FDA, Sept. 25, 2010: "Other than the copy of the IND exemption letter that was sent by Cornell as part of their package in response to

the OHRP inquiry, there are no known FDA records pertaining to the IND." (Obtained via FOIA.)

249 **suggesting that OHRP rely on his work:** "And thanks again for coming up with the plan for using the FDA memo"; Jerry Menikoff replying to Robert "Skip" Nelson, e-mail communication, Sept. 1, 2010 (obtained via FOIA).

249 **"ethically proper at every level":** Maria I. New, "Vindication of Prenatal Diagnosis and Treatment of Congenital Adrenal Hyperplasia with Low-Dose Dexamethasone," *American Journal of Bioethics* 10, no. 12 (2010): 67–68.

249 **the editors refused:** The exchange is reproduced at http://fetaldex.org/AJOB_Sept_2012.html.

249 **New is subject to no such limitation:** This was explained to me in e-mail correspondence with Robert "Skip" Nelson, May 24, 2011. See also Nelson through Murphy to Borror, Aug, 30, 2010.

250 **the "bad-ad" division of the FDA:** Alice Dreger to FDA Division of Drug Marketing, Advertising, and Communications, Sept. 10, 2010.

250 **They shut it down:** This news first came to us via a conversation between Ellen Feder and Svetlana Lajic (a member of the Swedish team), in an e-mail from Lajic to Feder, Sept. 19, 2011, subject "Question about your study of dexamethasone"; quoted in Dreger, Feder, and Tamar-Mattis, "Prenatal Dexamethasone," 285. The Swedes later published this information in Tatya Hirvikoski et al., "Prenatal Dexamethasone Treatment of Children at Risk for Congenital Adrenal Hyperplasia: The Swedish Experience and Standpoint," *Journal of Clinical Endocrinology and Metabolism* 97, no. 6 (2012): 1881–83.

250 **study of forty-three children:** Ibid.

251 **retrospective convenience-sample study:** Heino F. L. Meyer-Bahlburg et al., "Cognitive Outcome of Offspring from DexamethasoneTtreated Pregnancies at Risk for Congenital Adrenal Hyperplasia due to 21-Hydroxylase Deficiency," *European Journal of Endocrinology* 167 (2012): 103–10.

251 **highly skewed:** Analysis of this paper was provided in an epilogue to Dreger, Feder, and Tamar-Mattis, "Prenatal Dexamethasone."

251 **"prevents understanding of questionnaire":** Maria I. New, "Long-Term Outcome in Offspring and Mothers of Dexamethasone-Treated Pregnancies at Risk for Classical Congenital Adrenal Hyperplasia Owing to 21-Hydroxylase Deficiency," Rare Diseases Clinical Research Network, research protocol, (New York: Mount Sinai School of Medicine, 2007), 25 (obtained via FOIA).

251 **Maria New still does:** See "Prenatal Diagnosis and Treatment of Congenital Adrenal Hyperplasia," Maria New Children's Hormone Foundation, www.newchf.org/testing.php (accessed Aug. 1, 2014).

252 **The top hit:** Dreger, Feder, and Tamar-Mattis, "Prenatal Dexamethasone."

252 **The second hit:** Svetlana Lajic, "Prenatal Treatment of Congenital Adrenal Hyperplasia" (undated), at http://www.caresfoundation.org/productcart/pc/prenatal_treatment_cah.html.

252 **The third:** Alice Dreger, Anne Tamar-Mattis, and Ellen K. Feder, "Experimental Status of Prenatal Dexamethasone for CAH Re-Affirmed," *Endocrine Today*, October, 2011, at http://www.healio.com/endocrinology/news/print/endocrine-today/%7B547e98a4-7f10-495a-a01e-c5719cea8071%7D/experimental-status-of-prenatal-dexamethasone-for-cah-re-affirmed.

252 The fourth: Catherine Elton, "A Prenatal Treatment Raises Questions of Medical Ethics," *Time*, June 18, 2010.

252 The fifth: Tatya Hirvikoski et al., "Prenatal Dexamethasone Treatment of Children at Risk for Congenital Adrenal Hyperplasia: The Swedish Experience and Standpoint," *Journal of Clinical Endocrinology and Metabolism* 97, no. 6 (June 2012): 1881–83, doi:10.1210/jc.2012-1222.

252 The sixth: Alice Dreger, Ellen K. Feder, and Anne Tamar-Mattis, "Preventing Homosexuality (and Uppity Women) in the Womb?" *Bioethics Forum*, Hastings Center, June 29, 2010, www.thehastingscenter.org/Bioethicsforum/Post.aspx?id=4754&blogid=140.

252 The seventh: Alice Dreger, "IVF on Steroids: The Dangerous Off-Label Use of 'Dex' During Pregnancy," *The Atlantic* (January 13, 2013), at http://www.theatlantic.com/health/archive/2013/01/ivf-on-steroids-the-dangerous-off-label-use-of-dex-during-pregnancy/267187/.

CONCLUSION: TRUTH, JUSTICE, AND THE AMERICAN WAY

260 one tiny historical story: I am indebted to Charles Greifenstein of the American Philosophical Society for introducing me to this historical story.

EPILOGUE: POSTCARDS

264 *Journal of Urology*: Richard S. Hurwitz, "Long-Term Outcomes in Male Patients with Sex Development Disorders—How Are We Doing and How Can We Improve?," *Journal of Urology* 184, no. 3 (2010): 821–32.

264 Swiss National Advisory: Swiss National Advisory Commission on Biomedical Ethics, NEK-CNE, *On the Management of Differences of Sex Development: Ethical Issues Related to "Intersexuality"* (Berne: Nov. 2012), Opinion No. 20/2012, http://www.nek-cne.ch/fileadmin/nek-cne-dateien/Themen/Stellungnahmen/en/NEK_Intersexualitaet_En.pdf.

265 special rapporteur on torture: *Juan E. Méndez, Report of the Special Rapporteur on Torture and Other Cruel, Inhuman or Degrading Treatment or Punishment*, Human Rights Council of the United Nations, Feb. 1, 2013, www.ohchr.org/Documents/HRBodies/HRCouncil/RegularSession/Session22/A.HRC.22.53_English.pdf.

265 a recent study: Jürg C. Streuli et al., "Shaping Parents: Impact of Contrasting Professional Counseling on Parents' Decision Making for Children with Disorders of Sex Development," *Journal of Sexual Medicine* 10, no. 8 (Aug. 2013): 1953–60, doi: 10.1111/jsm.12214. See also Ellen K. Feder, *Making Sense of Intersex: Changing Ethical Perspectives in Biomedicine* (Bloomington: Indiana University Press, 2014).

265 through the American courts: M.C., a minor by and through his parents *Pamela Crawford and John Mark Crawford v. Dr. Ian Aaronson, Dr. James Amrhein, Dr. Yawappiagyei-Dankah, Kim Aydlette, Meredith Williams, etc.*, Civil Action No. 2:13-cv-01303-DCN (U.S. District Court for South Carolina, Charleston Division), filed May 14, 2013.

266 European clinicians again seem to be leading: Alice Dreger, "Gender Identity Disorder in Childhood: Inconclusive Advice to Parents," *Hastings Center Report* 39, no. 1 (Jan.–Feb. 2009): 26–29.

267 **"pink boys" and "blue girls":** See Alice Dreger, "Pink Boys: What's the Best Way to Raise Children Who Might Have Gender Identity Issues?," *Pacific Standard* (July 18, 2013), at http://www.psmag.com/culture/pink-boys-gender-identity-disorder-62782/.

267 **among transgender activists:** Zinnea Jones, "100-Plus Trans Women Stand Against Calpernia Addams and Andrea James: An Open Letter," *Huffington Post* (April 14, 2014), www.huffingtonpost.com/zinnia-jones/calpernia-addams-andrea-james_b_5146415.html.

268 **groundbreaking book:** Anne A. Lawrence, *Men Trapped in Men's Bodies: Narratives of Autogynephilic Transsexualism* (New York: Springer, 2013).

268 **made safer and better:** See, for example, Emily Newfield et al., "Female-to-Male Transgender Quality of Life," *Quality of Life Research* 15, no. 9 (Nov. 2006): 1447–57; Mohammad Hassan Murad et al., "Hormonal Therapy and Sex Reassignment: A Systematic Review and Meta-Analysis of Quality of Life and Psychosocial Outcomes," *Clinical Endocrinology* 72, no. 2 (Feb. 2010): 214–31; Tiffany A. Ainsworth and Jeffrey H. Spiegel, "Quality of Life of Individuals with and without Facial Feminization Surgery or Gender Reassignment Surgery," *Quality of Life Research* 19, no. 7 (Sept. 2010): 1019–24; and Anne A. Lawrence, "Factors Associated with Satisfaction or Regret Following Male-to-Female Sex Reassignment Surgery," *Archives of Sexual Behavior* 32, no. 4 (Aug. 2003): 299–315. See also Ray Blanchard, "The Case for Publicly Funded Transsexual Surgery," *Psychiatry Rounds* 4, no. 2 (Apr. 2000), 4–6, http://individual.utoronto.ca/james_cantor/index_files/Blanchard2000.pdf.

271 ***Noble Savages:*** Napoleon Chagnon, *Noble Savages: My Life Among Two Dangerous Tribes—the Yanomamö and the Anthropologists* (New York: Simon & Schuster, 2013).

271 ***Washington Post:*** Sebastiao Salgado, "The Yanomami: An Isolated Yet Imperiled Amazon Tribe," *The Washington Post*, July 25, 2014, at http://www.washingtonpost.com/wp-srv/special/world/yanomami/.

272 **ended up resigning:** David Cyranoski, "Controversial Bioethicist Quits Stem-Cell Company," *Nature* blog, Mar. 1, 2012, www.nature.com/news/controversial-bioethicist-quits-stem-cell-company-1.10151, doi:10.1038/nature.2012.10151.

272 **had mysteriously died:** Jiyeon Lee, "South Korean Company Cleared in Deaths Following Stem Cell Therapy," CNN, Dec. 14, 2010, www.cnn.com/2010/WORLD/asiapcf/12/14/south.korea.stem.cell.

272 **investigation by the FDA:** See the warning letter from FDA to Celltex Therapeutics Corporation, Sept. 24, 2012, www.fda.gov/ICECI/EnforcementActions/WarningLetters/2012/ucm323853.htm.

272 **Summer herself resigned:** See Christian Munthe, "Further on What's Cookin' at AJOB and Bioethics.net," *Philosophical Comment* blog Sept. 16, 2012, http://philosophicalcomment.blogspot.com/2012/09/further-on-whats-cookin-at-ajob-and.html.

272 ***AJOB* had essentially been forced to publish:** William Heisel, "Ethics Journal Corrects Record, Reveals Conflicts of Interest," *Reporting on Health*, Aug. 15, 2012, www.reportingonhealth.org/2012/08/14/ethics-journal-corrects-record-reveals-conflicts-interest.

273 **One recent laboratory study:** Marine Poulain et al., "Dexamethasone Induces Germ Cell Apoptosis in the Human Fetal Ovary," *Journal of Clinical Endocrinology & Metabolism* 97, no. 10 (Oct. 2012): e1890–97.

274 **study from Finland:** Natasha Khalife et al., "Prenatal Glucocorticoid Treatment and Later Mental Health in Children and Adolescents," *PLOS One* 8, no. 11 (Nov. 2013): e80194.

274 **Theresa Defino:** Theresa Defino, "Big Drop in OHRP Letters, Open Cases Raise Questions of Agency Commitment," *Report on Research Compliance* 8, no. 3 (Mar. 2011): 1–3, www.reportonresearchcompliance.com/rrc0311_reprint.pdf; and Theresa Defino, "'SUPPORT' Backlash Prompts Meeting, Guidance as Debate Moves Beyond OHRP" and "HHS Asked Oversight Agency, NIH to 'Align' About Disputed Study," *Report on Research Compliance*, vol. 10, no. 7 (July 2013): 1–6.

274 **only one investigation:** Theresa Defino, "With Just One Investigation in 2013, OHRP Seems 'Invisible' After SUPPORT Dust-Up," *Report on Research Compliance*, May 2014, www.reportonresearchcompliance.com/rrc-reprint-0514.pdf.

275 **Kari Christianson:** quoted in Alice Dreger, "IVF on Steroids: The Dangerous Off-Label Use of 'Dex' During Pregnancy," *The Atlantic*, Jan. 16, 2013, www.theatlantic.com/health/archive/2013/01/ivf-on-steroids-the-dangerous-off-label-use-of-dex-during-pregnancy/267187.

276 **the Nazis:** I will forever be grateful to Ellen Feder for our long discussions of Hannah Arendt during the years we lived together with dex.

INDEX

Page numbers beginning with 281 refer to endnotes.